Wind Energy

Recent Titles in the

CONTEMPORARY WORLD ISSUES

Series

Entertainment Industry: A Reference Handbook
Michael J. Haupert

World Energy Crisis: A Reference Handbook
David E. Newton

Military Robots and Drones: A Reference Handbook
Paul J. Springer

Marijuana: A Reference Handbook
David E. Newton

Religious Nationalism: A Reference Handbook
Atalia Omer and Jason A. Springs

The Rising Costs of Higher Education: A Reference Handbook
John R. Thelin

Vaccination Controversies: A Reference Handbook
David E. Newton

The Animal Experimentation Debate: A Reference Handbook
David E. Newton

Steroids and Doping in Sports: A Reference Handbook
David E. Newton

Internet Censorship: A Reference Handbook
Bernadette H. Schell

School Violence: A Reference Handbook, Second Edition
Laura L. Finley

GMO Food: A Reference Handbook
David E. Newton

Books in the **Contemporary World Issues** series address vital issues in today's society such as genetic engineering, pollution, and biodiversity. Written by professional writers, scholars, and nonacademic experts, these books are authoritative, clearly written, up to date, and objective. They provide a good starting point for research by high school and college students, scholars, and general readers as well as by legislators, businesspeople, activists, and others.

Each book, carefully organized and easy to use, contains an overview of the subject, a detailed chronology, biographical sketches, facts and data and/or documents and other primary source material, a forum of authoritative perspective essays, annotated lists of print and nonprint resources, and an index. Readers of books in the Contemporary World Issues series will find the information they need in order to have a better understanding of the social, political, environmental, and economic issues facing the world today.

CONTEMPORARY WORLD ISSUES

Wind Energy

A REFERENCE HANDBOOK

David E. Newton

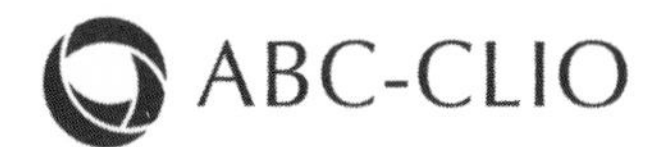

Santa Barbara, California • Denver, Colorado • Oxford, England

Library of Congress Cataloging-in-Publication Data

Newton, David E.
Wind energy : a reference handbook / David E. Newton.
pages cm. — (Contemporary world issues)
Includes bibliographical references and index.
ISBN 978-1-61069-689-0 (hard copy : alk. paper) —
ISBN 978-1-61069-690-6 (ebook) 1. Wind power. 2. Wind power—History. 3. Electrical engineers—Biography. I. Title.
TJ820.N49 2015
333.9'2—dc23 2014027127

ISBN: 978-1-61069-689-0
EISBN: 978-1-61069-690-6

19 18 17 16 15 1 2 3 4 5

This book is also available on the World Wide Web as an eBook. Visit www.abc-clio.com for details.

ABC-CLIO, LLC
130 Cremona Drive, P.O. Box 1911
Santa Barbara, California 93116-1911

This book is printed on acid-free paper ∞

Manufactured in the United States of America

Contents

List of Tables

Preface

The world is passing through a historical period that has sometimes been called the Age of Fossil Fuels. That term is based on the fact that humans currently depend heavily on coal, oil, and natural gas to meet most of their basic needs. These fuels power most forms of transportation, such as cars, trucks, airplanes, and trains. They are used to operate most industrial operations, from the production of a host of commercial products to the generation of electricity. They are the source of the heat used for residential and commercial buildings, as well as the heat needed for chemical and industrial processes. And they are the raw materials from which a countless number of personal and commercial products are made, ranging from drugs and pharmaceuticals to cosmetics to many kinds of plastics.

But many energy experts are convinced that the world's supply of fossil fuels will not last forever. Because of the way by which coal, oil, and natural gas are produced, they will eventually be available in amounts far less than may be needed to continue meeting this wide variety of demands. A good deal of controversy exists as to when the supply of these materials will have reached a point of peak production; most authorities argue that the question is not *if* the world will run out of fossil fuels but *when.* Where do humans then turn for the energy resources they need to maintain the high level of civilization reached in this century?

The answer to that question for many people is renewable energy sources, forms of energy that occur naturally and are available in abundant supplies that should continue to be available essentially forever. The renewable energy sources that have drawn the most attention from researchers and policy-makers in the last half century (at least) have been solar power, hydroelectric power, geothermal energy, energy from biomass, and wind power. This book reviews the role of wind power in human history, its potential to be a major source of energy in the near and more distant future, and the questions, problems, and issues associated with the development of wind energy.

Humans have used wind energy for a variety of purposes for more than two millennia. For most of that long period, the primary role of wind power was the operation of sailing ships, the lifting of water (usually for irrigation systems), and the grinding of grains. It was not until the end of the nineteenth century that inventors first began to explore the possibility of using wind energy for the production of energy. At that point in history, major modifications in the historic windmill began to appear. Instead of building a machine (the windmill) that could provide for the needs of a single family, a small farm, or a modest community, inventors began to find modifications by which a wind device could generate electricity for large numbers of people, families, businesses, and communities. This change marked the beginning of the era of wind turbines and wind farms.

For much of its modern history, the story of wind power has been one marked by ups and downs in popularity, changes dependent to a large extent on the status of fossil fuel prices. During times when the price of fossil fuels increased significantly (as during the Arab Oil Embargo of 1975), interest in wind power increased correspondingly. During times when the price of fossil fuels dropped, interest in wind power diminished, because it could not compete economically with fossil fuels for most applications. It is hardly surprising, then, that private companies have been slow to become involved in research and development of wind machines and/or to take the initiative in building large wind facilities. For most of the last half century (at least), developments in wind power technology have depended almost entirely on favorable governmental policies at the local, state, and national levels. For example, without federal and state tax credits and other incentives, a much smaller number of wind turbines and wind farms would probably have been built in the United States or most other countries of the world. One of the fundamental questions about wind power, then, is what can be done in the future not only to continue governmental support of wind technology and development, but also what can be done to encourage private enterprise to take a larger role in the promotion of wind energy in the energy equations of the United States and other nations around the world.

In the second decade of the twenty-first century, wind power has become a very small, but not unimportant, part of the energy program of many developing nations. In some countries, such as the United States, China, Denmark, and the United Kingdom, that statement is more true than it is in other countries. In many nations where wind power might be expected to be (or to become) a major supplier of electricity, a very different story is true. Very few developing nations, for example, have yet to develop wind systems that make even a minor contribution to their energy needs.

This book opens with a chapter that provides a historical background of the way wind power has been used by humans over the past 2,000 years. The narrative then shifts to the modern day and a review of the questions, problems, and issues associated with the development of wind energy systems in the United States and other nations. Some of the ways in which these problems are being addressed and some solutions for the future are also discussed. Chapter 3 of the book provides an opportunity for stakeholders in the field of wind energy to present their views on important specific issues in the field. The remainder of the book provides a number of resources—an annotated bibliography, chronology, glossary, and list of important individuals and organizations in the field—of value in conducting further research on this topic.

Wind Energy

1 Background and History

Knowledge and use of wind power date to the very earliest period of human history. That fact should hardly be surprising. Everyone knows that a slight breeze can lift a leaf or bend a tree branch. Stronger winds can uproot a tree or blow the roof off a structure. No one should be surprised that the earliest humans observed the power of the wind and found ways to put that energy to work for them.

Early Sailing Ships

Archaeologists now believe that the earliest known use of wind power by humans can probably be traced to the Neolithic Period, which began sometime around 10000 BCE. During that era, humans apparently learned how to make simple sails out of animal hides or woven cloth. These sails were used to power the first boats when the wind was available. In fact, some researchers have even argued that the North American continent was first discovered and inhabited more than 16,000 years ago by European sailors, who traveled across the Atlantic Ocean in just such boats (Stanford and Bradley 2012).

Much of what scientists know about very early sailing ships comes from the art, architecture, and literature of ancient

---

One of Christopher Columbus's sailing ships: the *Niña*. (Library of Congress)

Egypt. Drawings, sculptures, and, in some cases, actual physical remains give researchers a good idea as to what early Egyptian sailing ships looked like. These first boats were relatively simple in that they needed essentially no external source of power to travel down (north) on the Nile River, and only the simplest external power source in going up (south) against the tide on the river. That external source of power might be the work of human rowers or, when the wind was blowing, sails designed to capture and use wind power to make their trip upstream. The Greek historian Herodotus (484–425 BCE) provides one of the earliest written descriptions of these Egyptian boats. He describes in detail the way in which such boats were built, adding that

> they have a mast of acacia and sails of papyrus. These boats cannot sail up the river unless there be a very fresh wind blowing, but are towed from the shore. ("Herodotus: The Second Book of the Histories, Called Euterpe" 2014)

In an extraordinary bit of good fortune, historians actually have physical examples of the types of sailing ships used in ancient Egypt. In 1954, for example, archaeologists found an essentially intact ship built in about 2500 BCE for the pharaoh Khufu. The ship was apparently buried in a pit adjacent to one of the great pyramids near Giza and is now housed in a dedicated building close to its discovery site ("Explore a Pharaoh's Boat" 2009; for an excellent overview of sailing ships technology in ancient Egypt, see Holmes 1906 or Chatterton 2010).

The primary purpose of sailing ships during the earliest period of Egyptian history appears to have been the transport of materials and completed structures along the Nile River. For example, the pharaoh Merenre Nemtyemsaf I (reigned from 2283 to 2278 BCE) appears to have used sailing ships to collect materials and monuments for the construction of his burial place. Commander of the expeditions to transport those items later wrote:

> His Holiness, the King Mer-en-Ra, sent me to the country of Abhat [in modern day Sudan] to bring back a sarcophagus with its cover, also a small pyramid, and a statue of the King Mer-en-Ra, whose pyramid is called Kha-nofer ("the beautiful rising"). And his Holiness sent me to the city of Elephantine to bring back a holy shrine, with its base of hard granite, and the doorposts and cornices of the same granite, and also to bring back the granite posts and thresholds for the temple opposite to the pyramid Kha-nofer, of King Mer-en-Ra. (Holmes 1906)

Some researchers have credited the ancient Egyptians with an even more sophisticated understanding of wind power. Maureen Clemmons, who holds a doctorate in organization change, became convinced in the late 1990s that the ancient Egyptians may have harnessed wind power to lift heavy objects in their construction of the pyramids. After studying the technical issues involved, she devised a scheme to replicate that process and, with the aid of a team of designers and engineers, she tested her theory in 2000. The team successfully lifted a 25-ton obelisk, suggesting that the Egyptians may, in fact, have found a way to use wind power to move massive objects in their construction projects. (See Clemmons and Cray 2001; it should be noted that most archaeologists have not been convinced by this demonstration that wind power was used for such purposes by the ancient Egyptians.)

In any case, the use of sailing ships for the transport of cargo eventually began to spread well beyond Egypt into the Mediterranean and other parts of maritime Europe. These voyages were necessitated by the demand in Egypt for raw materials that were not available in that country. They headed northeast, for example, to the eastern shores of the Mediterranean, in modern Lebanon, to collect timber, and they went southeast along the eastern coast of Africa to the ancient land of Punt, to collect incense and spices that were essential for traditional religious ceremonies. The earliest written record of these expeditions dates to the reign of King Sahure (reigned 2487 to 2476 BCE) and describes the transport of "80,000 measures of myrrh, [6,0001—of electrum, 2,600 [-] staves" (inability to translate this text in full; see Breasted 1906).

Space constraint does not permit a detailed discussion of the development of wind power for sailing ships over the centuries. Some of the most important developments in that history are summarized here.

Second century CE Invention of the triangular, or lateen, sail. Prior to this time, the square sail was by far the most common type of sail used in the Western world. The lateen was adopted because it has advantages over the square sail in certain types of wind situations (Whitewright 2009).

ca. 1000 CE Norwegian explorer Leif Eriksson conducts one of the first grand voyages of discovery in a sailing ship of a type called the knarr. On his voyage, Eriksson is thought to have been the first European to land in the New World (Carter 2000).

ca. 1400 CE A new type of sail rigging, called fore-and-aft, is used for the design of sailing ships that became known as barques (barks). The design introduced a new orientation for at least one or more of the sails, running longitudinal to the boat's keel, rather than perpendicular to it (Chatterton 1912).

Sixteenth century A proliferation of sailing ships of many sizes, shapes, and configurations begins to flood the world's waters. For the next three centuries, countless numbers of baleners, barques and barquentines, bilanders, brigs and brigantines, caravels, clippers, corvettes, East Indiamen, frigates, galleons, men of war, packets, schooners, sloops, tartanes, windjammers, yachts, and other sailing vessels carried cargo and passengers, fought battles, carried explorers and researchers, and played many other roles in the world's economy. The Age of Sailing

Ships (sometimes called the Golden Age of Sailing Ships) is sometimes said to have begun with the Battle of Lepanto in 1571, the last naval skirmish in which oared-boats played a significant role in the Civil War Battle of Hampton Roads in 1862, when the Confederate States' steam-powered CSS Virginia destroyed two of the North's largest sailing ships—the USS Cumberland and the USS Congress. (A number of excellent resources are available for more detail on ships in action during this period. See, for example, Ross 2013; Smith 2009.)

1720 The Water Club of the Harbour of Cork (Ireland) is formed to provide for sporting races of sailing ships among private individuals. Prior to this time, sailing ships had been used almost exclusively for the transport of goods and passengers, for exploration, and for military activities. The club is now known as the Royal Cork Yacht Club, and its claim to being the first such club in the world is challenged by the Neva Yacht Club of St. Petersburg, Russia ("Club History" 2014).

1851 The first America's Cup race is held around the Isle of Wight. The race becomes the most famous test of sailing ships in modern history between the fastest-sailing yachts in the world. The thirty-fifth race was historic when the required design for contesting yachts made use of the most modern technology available in sailing ship construction, all but eliminating a small handful of craft capable of competing in the event ("America's Cup: History" 2014).

1902 The Thomas W. Larson was launched at the Quincy (Massachusetts) shipyards. It was the largest pure sailing ship ever built, with a steel hull and twenty-six sails attached to its seven masts (Thomas W. Lawson 2014).

Wind Power at Work: The Beginnings

In addition to the invention and development of sailing ships, humans have used wind power for two other purposes for at least 2,000 years. One of the most commonly cited of such uses dates to the seventeenth century BCE, when the Babylonian emperor Hammurabi (reigned 1792 to 1750 BCE) is said to have ordered the construction of a system for bringing water from natural rivers to farms through a complex system of irrigation ditches. The system was designed to use a primitive form of windmills to scoop water out of the rivers and into the irrigation ditches. (This attribution is often questioned, and it seems to stand out as a rare mention from a period that early in history. See Wood, Ruff, and Richardson 2014; the primary source for this attribution is Golding 1976, 7, citing Flettner 1926, 94.)

Histories of the use of wind as a power source often suggest that the first windmills were constructed and used in East Asia, usually in Afghanistan, China, India, Tibet, or, most commonly, Persia, sometime prior to about 500 CE. Documentation for this assertion is generally absent or questionable. A number of scholars have reviewed putative evidence for the existence of windmills during this period and have generally found that evidence to be unreliable because of "mistranslations, revisions, and interpolations by other hands over the centuries" (Shepherd 1990, 4). In fact, much of this "evidence" actually consists of forgeries or other changes made "at the whims of revisionists" (Shepherd 1990, 4).

Most scholars do agree, however, that windmills were in wide use throughout Persia in the period between the sixth and tenth centuries CE ("Illustrated History of Wind Power Development" 2014). One of the oldest manuscripts describing the use of wind power can be found in the manuscript *Kitab nukhbat al-dahr fi ajaib al-barr wal bahr* (*Selection of the Age on the Wonders of the Land and the Sea*), by the thirteenth-century Arabic geographer Shams al-Din al-Ansari al-Dimashqi. The most popular windmills consisted of a vertical axis to which were attached vertical sails attached to the axis by horizontal struts in a style that came to be known as a panemone design (al-Hasan and Hill 1986, 55). Some authorities believe that this design was adapted from that used in water-driven mills for use in more arid regions, where running water was not available (Forbes 1965, 119). The panemone design is not a very efficient system for capturing wind energy, but it is simple to build and continues to be used in today's world ("Illustrated History of Wind Power Development 2014"; a video showing the operation of a modern panemone is available at http://www.ztopics.com/Panemone%20windmill/).

There are good reasons to believe that windmills were also being used in China prior to, during, or shortly after their adoption in Persia. One reason for the uncertainty about this point is that historians of science have generally devoted less research to historical developments in China and other parts of East Asia than they have to the Middle East and Europe. In the view of the preeminent historian of Chinese science, Joseph Needham, the windmill may have been introduced to China as early as the Sung dynasty (960–1279 CE) or as late as the Yuan dynasty (1271–1368) (Needham 1965, 561). Needham and other scholars have been convinced that the earliest written record of windmills in China dates to 1219 CE, when the great statesman Yehlu Chhu-Tshai wrote about such devices with a design identical to those then being used in Persia (Ronan 1994, 274).

Although a great deal of conjecture is involved in dealing with the oldest windmills, it appears that they almost certainly were used for two major purposes: the movement of water, as in filling irrigation ditches, and grinding grain. For the latter purpose, a number of modifications were made in the simplest panemone design. For example, the typical Persian windmill used for grinding grain appears to have been two stories tall, with the grinding wheel on the top floor and a second millstone attached to the rotating axle on the lower floor. Adjustments were made in the sails themselves, or a surrounding structure was built around the windmill to control the amount of wind admitted to the sails. Such caution was necessary because if the sails turned too rapidly, the grinding wheel would also turn so fast as to scorch the grain (Derry and Williams 1993, 254).

Windmills in Europe

As with many questions about the early history of wind power, the issue as to when and how windmills came to Europe is the subject of some debate. A number of popular articles and Web sites mention a grant of land given in 1105 to the abbot of Savigny in France for the construction of a windmill (see, for example, "Energy Time Machine" 2014). Such claims have long been rejected, however, for a variety of reasons (see, especially, Delisle 1851; Kealey 1987, 12). Most authorities now seem to agree that the first certain mention of windmills in Europe comes in a deed issued in 1180 allowing the construction of such a machine in the canton of Sauveur-le-Vicomte in France. The first windmill in Portugal was apparently built two years later, and in England, in 1190 (Kealey 1987, 12).

European windmills were built on a quite different principle than were those in Asia. Instead of mounting the windmill sails on a vertical axis, they were placed on a horizontal axis, parallel to the ground. The axis itself was then supported by a vertical pole or other means of support. The horizontal windmill has some distinct advantages (as well as some disadvantages) compared to the vertical axis machine. In the latter case, the force of the wind is able to drive the sails only half of the time. Those sails on the side of the axis away from the wind do not, of course, feel the wind's impulse and so contribute nothing to propelling the rotating wheel. In fact, their mass acts as a drag on the system, and they actually reduce the net power produced by the rotating wheel. The horizontal wheel does not have this disadvantage, since the face of the sails is directed toward the wind at all times, so the full force of the wind is captured in the movement of the wheel. (For a video comparison of two simple examples of vertical and horizontal windmills, see "Vertical Wind Turbine v Horizontal Wind Turbine" 2014.)

By the beginning of the fourteenth century, interest in the use of windmills for a variety of purposes was growing rapidly. In Great Britain, from which perhaps the best statistics are available, as many as 4,000 operating windmills were in use in 1300, a number that was to grow to at least 10,000 a century later (Ambler and Langdon 1994, 5). The windmills were most commonly built in areas where there were no or only slow-running rivers, where traditional water power systems could not be used. The primary purpose of the English windmills was to provide the power needed to grind corn and to prepare cloth, a process known as *fulling*. Apparently, they were not employed for the movement of water or for any other purpose as they were in some other countries at the time (Ambler and Langdon 1994, 14).

An important example of the alternative uses of windmills can be found in the region sometimes known as the Lowlands, later the Dutch Republic or United Provinces and, now, the Netherlands. Records exist of windmills in the region as early as 1200, but these devices were apparently used only for the grinding of corn. As the region began to develop a program of draining wet and damp regions to extend its land area, developers turned more and more toward the use of windmills for that purpose. The first confirmed use of windmills for draining land dates to 1414 in the town of Reijerwaard. As the land reclamation program progressed, more and more windmills were built to carry out the transfer of water until an estimated 200 windmills were in operation at the beginning of the sixteenth century (Berend 2013, 32), a number that was to rise to 700 windmills by the end of the seventeenth century and to a maximum of 900 prior to the conversion of pumping stations to steam power in the eighteenth century (Derry and Williams 1993, 256–257; for an excellent overview of the status of wind power during this period, see Stokhuyzen 1965, Chapter 1).

Interestingly, as the number of windmills increased, they began to find a wider range of uses. In the Lowlands, for example, windmills were increasingly used as timber sawmills (the first appearing in 1592), as oil mills (for the production of oils from seeds, nuts, and grains, 1582), and as paper mills (1586) (Stokhuyzen 1965). Additional uses included the grinding of powders to make dyes, the grinding of spices, the provision of ventilation systems for mines, the production of gunpowder, and the manufacture of snuff powder (Ragheb 2014).

Development in Wind Power Technology

The growth in use of wind power throughout Europe after the twelfth century was encouraged to a large extent by developments in windmill technology in response to a number of fundamental shortcomings with simple horizontal- and vertical-axis devices. The basic problem in any kind of windmill, of course, is to keep the sails faced toward the wind. As the direction of the wind changes, so must the orientation of the windmill. In its earliest form, the horizontal windmill design solved this problem by mounting the machinery of the windmill, called the *house* or the *buck*, around a central pole, usually made of oak, supported by four angled posts anchored to the ground. This form of windmill became known as a *post windmill.* (For diagrams of various types of windmills, see Stokhuyzen 1965, Chapter 2.) The house was then rotated around the central pole so that the sails always faced into the wind. (Gasch and Twele 2012, 19–20; a number of videos showing the operation of a post windmill are available on YouTube as, for example, https://www.youtube.com/watch?v=24uTsPfI2dA.)

A popular modification of the post mill was produced by hollowing out the central support post, allowing the machine's drive shaft to be fitted into the center of the post. This design became known as the *hollow-post mill.* This form of the windmill became especially popular in the Lowlands, where it was used for draining lakes, ponds, and swamps, and was known as the *wipmolen.*

Another modification of the basic horizontal windmill was developed because of the practical problems of rotating the large and bulky house around the center post. In this modification, the house itself remained fixed in position around the post, and only the cap on the house rotated. This design became known as a *tower windmill.* Tower mills were more expensive to build because they were more complicated than post mills, but they were also more efficient because they could be made larger with larger sails that could function in reduced wind speeds. (Stokhuyzen 1965 [Chapter 2] provides much more detail on a greater variety of modifications of the basic post and tower windmills.)

A final windmill design worthy of mention is the so-called *smock mill,* a variation of the tower mill in which the walls of the mill are made of a lighter framework than that used in the tower mill. That framework may consist of wood, tar paper, slate, or sheet metal and is generally less expensive to build and repair than its tower cousin. (See, especially, Duell 2014 for a detailed look at the construction of a smock mill.)

In addition to modifications in the basic design of windmills, a number of changes in mill accessories improved the efficiency of the machines. Perhaps the most important of these adaptations was the invention of the fan-tail (or fantail) in 1745, by English millwright Edmund Lee. A fantail is a small windmill mounted at right angles to the sails of a windmill opposite the main sails. When the wind blows directly at the fantail, it does not move. When the wind shifts direction and blows at an angle to the fantail, it moves the fantail into a new position such that it again lines up directly with the wind. When the fantail moves, it also moves a connecting rod that joins the fantail to the main sail mechanism of the mill. With this invention, human intervention was not required to make sure that the mill sails were always pointed directly at the wind ("Wind Powered Factories 2009").

Windmills Arrive in North America

Early settlers in North America brought with them the windmill technology with which they were most familiar in their home countries. Dutch settlers in New Amsterdam (modern-day New York City) brought designs for their customary tower windmills, whereas English settlers in other parts of the colonies (especially at first in Jamestown, Virginia) constructed post mills. Generally speaking, however, windmills were not a major source of power in the early colonies (Morrison 1952, Chapter 3, "New England Colonial Architecture").

Such was not the case in the western part of the continent, however. As explorers and pioneers began moving westward in the early 1800s, they found that a major hindrance to the development of land for farming and pasturing was the lack of water. In many locations, the only hope for establishing communities was finding underground water reservoirs and devising methods for bringing water to the surface. Coincidentally, demands created by the opening of the West itself, primarily the development of a massive railroad system, created another major source of demand for water collection systems.

The solution to this problem was the development of an entirely original type of windmill, sometimes known as the *wind engine*, or simply the *American windmill.* The wind engine was invented in 1857 by Connecticut mechanic Daniel Halladay. The idea for a new type of windmill was not actually Halladay's but a suggestion from one of his colleagues, John Burnham. Burnham had become interested in designing, building, and selling water pumps for use in his native New England. He became convinced that harnessing the wind would be an ideal way to power such devices. Lacking the technical skills himself to design such a device, however, he encouraged Halladay to work on the problem.

The design that Halladay produced is perhaps the simplest of all windmills popular in modern history. The wind engine consisted simply of a sturdy wooden pole set upright into the ground with a rotating cap at the top. Attached to the cap were a number of sails, most commonly at first, four, or six sails. Attached to the rear of the cap was a fantail that allowed the cap to adjust its position so that the sails were always facing into the wind (Baker 1985, Chapter 2; "History" 2014). When the Halladay model went into production, it took a number of somewhat different forms. For example, the blades could be anywhere from 6 feet in diameter in the smallest models to more than 16 feet in the largest models. Some models came with sails made of cloth mounted on steel frames, whereas others had thin wooden blades.

Another important adaptation was the attachment of sails or blades to the supporting structure by pivots, allowing them to shift their orientation to the wind. This adaptation was important because all windmills operate with differing efficiency depending on the strength of the wind. While a strong wind is useful in developing maximum wind power, excessive wind can cause damage to the sails or the supporting structure. For this reason, windmill designers in both Europe and North America had long worked on methods to adjust the amount of sail or blade surface exposed to the wind. Halladay's adaptation eventually became known as the *sectional wheel* because the wooden blades that it contained could be pivoted so as to expose more or less surface area to the wind (Baker 1985, Chapter 2; "History" 2014).

Halladay's invention was by no means the only attempt to solve the problem of bringing water to the West. A second important development was the solid-wheel mill. This design was developed by Rev. Leonard H. Wheeler, at the time a missionary to the Ojibwa Indian tribe in Wisconsin. Wheeler's design was motivated to some extent by the loss of a traditional wind energy device he was using to pump water during a severe windstorm. Wheeler determined that he find a way to build a windmill that could adjust itself when winds became too severe. The blade-pivoting arrangement on the Halladay mill had obviously not been adequate to deal with the windstorms Wheeler's windmill had experienced!

His solution to the problem, which eventually became known as the Eclipse windmill, was to mount a second fantail at the rear of the mill paddles. The second fantail was placed in such a way that heavy winds caused it to turn away from the wind direction, carrying the paddles or sails with it and placing them parallel to the wind direction. When the wind died down, a weight suspended from the second fantail returned the wheel to its original position. (Diagrams and descriptions of this model are available at Baker 1985, Chapter 2, and "History" 2014.)

A third step forward in the development of the American windmill was the invention in 1888 of a device known as the Aermotor by inventor LaVerne Noyes and engineer Thomas Perry. The Aermotor possessed no fundamentally new design features to set it apart from the Halladay or Wheeler machines. Its unique features involved the use of the strongest, sturdiest, and lightest building materials available at the time (sheet metal and steel), along with the application of engineering principles that allowed the mill wheel to rotate and transmit its kinetic energy in the most efficient method possible.

The Aermotor design experienced almost instantaneous success. In the first year of operation, the Aermotor Company sold only 45 units, but that number climbed to 2,288 in 1889, 6,268 in 1890, and 20,049 in 1891. By the end of the century, the company had claimed half the market for windmills in the United States, with total sales of more than 800,000 units (Baker 1985, 38).

The variety of innovative devices invented by Halladay, Wheeler, Noyes, Perry, and their compatriots assured a widespread and ongoing success for the American windmill in the West. By 1888, there were more than seventy-seven companies devoted to the manufacture of windmills, and only a decade later, over the next century, more than six million such devices were installed, the vast majority on farms, where they were used to pump water, grind grain, operate grain elevators, card wool, and process wheat and corn (Asmus 2001, 31; "History" 2014; "Illustrated History of Wind Power Production" 2014).

The First Windmill Dynamos

At almost the same time that LaVerne Noyes and Thomas Perry were developing the Aermotor windmill, another important step forward in the use of wind power was taking place in Cleveland, Ohio. Inventor Charles F. Brush built a giant windmill with a rotor 56 feet in diameter attached to a steel tower weighing more than 40 tons and having a height of 60 feet in the backyard of his home on Euclid Avenue. The rotor consisted of 144 blades with a total surface area of 1,800 square feet to which was attached a fantail that was 60 feet long and 20 feet wide. In operation, the rotor spun at a speed of 500 revolutions per minute.

The rotation of the windmill was transferred by a series of pulleys and belts to an electrical generator that was used to charge a bank of more than 400 batteries in the basement of the house. These batteries were used to operate 350 incandescent lamps, three motors, and two arc lights with a maximum power production of 12 kilowatts ("Charles F. Brush" 2014; "Mr. Brush's Windmill Dynamo 1890," 54). The windmill survived until 1909, at which time electricity from fossil fuels was becoming more readily and less expensively available.

In retrospect, it is difficult to overestimate the potential impact of this invention, the creation of one of the world's first windmill dynamos. (The term *dynamo* refers to any device that converts mechanical energy into electrical energy.) One observer of the time assessed the potential of the invention in this way:

> Occasionally a new invention will appear which will gradually affect a whole range of allied inventions and industries in such a way as to entirely change time-honoured customs, inaugurate new practices, and establish new arts. The commercial development of [Mr. Brush's invention] is a notable example of this. ("Mr. Brush's Windmill Dynamo 1891," 14)

In fact, Brush's invention was not precisely the first example of wind-powered electrical generation. About a year earlier, Professor James Blyth of Anderson's College, in Glasgow, Scotland, invented a similar device for the electrification of his home at Marykirk in Kincardineshire. Blyth's windmill was more modest than Brush's later machine. It was a horizontal-axis device with a diameter of 33 feet and vertical sails made of cloth fabric. (For a photograph of the machine, see http://tmblr.co/Zx2_Yp1Fes_rU.) Blyth received a patent for his invention and read a paper before the Royal Society of Edinburgh about it. His offer to provide electricity from the machine to local agencies, including the Royal Lunatic Asylum, Infirmary, and Dispensary of Montrose, however, was declined, to a large extent because religious leaders saw the device as an "engine of the devil" (Price 2005, 2014).

The Brush windmill dynamo opened a new age for wind power in the United States and, ultimately, the rest of the world. Inventors soon realized, however, that the Brush model was far too bulky and expensive to be used for many of the situations in which wind-powered electricity would be useful and convenient. No one reached this realization and solved that problem more quickly than did the Jacobs brothers, Marcellus (usually known as M. L.) and Joe. After moving from their home in Indiana to the remote town of Vida, Montana, they set about installing the most modern electrical equipment for both ranch operations and home appliances. The electrical systems that worked well in Indiana, however—such as gasoline-powered appliances—were out of the question in Vida. The nearest gasoline supplies were a three-day round trip, far too great an inconvenience to be practical. So the Jacobs brothers decided to design a simple wind-powered machine that would provide the electricity they needed at lesser cost and greater convenience.

That device was one of the world's first commercially successful wind turbines, a horizontal-axis "windmill" consisting of three propeller blades 14 feet in diameter connected directly to an electrical generator. The Jacobs machine could be considered a windmill only in the most general sense and is more properly called a wind *turbine* because it is designed to convert the mechanical energy of moving air directly into electrical energy. In fact, the Jacobs wind turbine looked and behaved much like a modern airplane propeller, hardly a surprise considering M. L.'s interest in, and growing experience with, aircraft engines and the principles of aerodynamics ("Beginnings," 2014; for diagrams and photographs of early Jacobs wind turbines, see "The Jacobs Wind Electricity Company" 2014).

The Jacobs brothers experienced almost immediate commercial success as their neighbors quickly realized the convenience and power of their machines. They became incorporated as the Jacobs Wind Electricity Company in 1928 and, four years later, moved their operations back to the Midwest in Minneapolis and later, Minnetonka, Minnesota, where the company continues to operate today ("Beginnings" 2014).

Decline of the Wind Industry in America

In some ways, the rise of the wind industry in the United States after World War I occurred at an inopportune time. True, the production and sale of windmills experienced a spurt in growth toward the end of the 1920s and the beginning of the 1930s, as shown in Table 1.1. But that growth was to be short-lived, and both production and sales began to drop off fairly quickly. Production of windmills that reached the hundreds of thousands in the late 1920s dropped to a handful—fewer than 10,000 annually—by the mid-1950s.

A number of factors were responsible for this decline. One factor was the onset of the Great Depression, a period of worldwide economic downturn that began in about 1930. This economic disaster coincided in the American Midwest, home to a vast number of windmills, with the Dust Bowl, one of the greatest ecological disasters in American history. These two

Table 1.1 Production of Windmills, 1927–1963

Year	Production (units)
1927	120,586
1928	149,923
1929	137,166
1930	94,061
1931	46,663
1935	69,944
1936	94,637
1937	87,713
1938	63,444
1939	63,098
1954	11,289
1958	6,983
1963	7,562

Source: All data are from the *Statistical Abstract of the United States* of various years. Complete data are not available from this source because various editions of the document reported data about windmills in a variety of forms (such as the financial value of windmills rather than the number produced) or reported no data at all, as was especially the case as interest in windmills declined.

events combined to create financial hardships for residents of the Midwest that made the purchase, operation, and repair of windmills a major economic burden. A number of windmill companies attempted to survive this period by offering special sales on their machines, providing financing plans, and offering assistance to users in other ways. But these efforts were largely ineffective within the much larger context on national and international economic crisis (Baker 1985, 106–108). Ironically, the windmill business in the American Midwest began to fail almost as soon as it began to grow and become an integral part of American agriculture in the region.

Symbolic of this trend was the failure in 1956 of the Jacobs Wind Electric Company, which blamed not only poor economic conditions but also a second factor—rural electrification. By the 1930s, the vast majority of urban dwellers in the United States had access to electricity. However, such was also not the case in rural areas, where perhaps no more than one in ten residents had electrical service. One of the first acts in President Franklin Delano Roosevelt's New Deal program was his issuance of Executive Order 7037 on May 11, 1935, creating the Rural Electrification Administration (REA) ("The Politics of the REA" 2014; "Rural Electrification" 2014). The REA was to be a federal agency with authority to provide loans to cooperatives organized to establish electrical grids to rural areas. Within a relatively short period of time, rural residents who had previously had to rely on windmills as one way of obtaining electricity now had access to power grids similar to those in urban areas. While a great boon to rural areas in many ways, the REA also turned out to be one more nail in the coffin of wind power in the Midwest (Baker 1985, 107).

A third factor in the decline in dependence on windmill power was the growth in production and consumption of fossil fuels—coal, oil, and natural gas—in the United States and around the world. Table 1.2 provides an overview of this growth, which is perhaps more clearly shown in a graph such as the one on which these data are based. (See the source given for data in Table 1.2.) The relatively low cost of most fossil fuels, such as crude oil, was also a factor. (See also Table 1.3 for historic trends in the cost of crude oil.) As Table 1.3 shows, crude oil (and therefore gasoline, diesel oil, kerosene, and other products made from it) was relatively inexpensive with little or no increase in price from the mid-1920s to the mid-1970s. Throughout that period, consumers could depend on an abundant supply of fossil fuels at low cost, with which to operate almost every imaginable mechanical device, providing a ready alternative for rural residents to the windmill. As historian of American windmills, T. Lindsay Baker, has written, the competition between internal combustion engines and windmills for use as a source of power on farms existed as early as the 1890s, but combined with factors such as the economic downturn and rural electrification in the 1930s, "the demand for windmills changed more dramatically beginning in 1935" (Baker 2007, 29).

Table 1.2 Oil Consumption in the United States, 1915–2010

Year	Consumption (quadrillion Btu)
1915	1.4
1920	2.7
1925	4.3
1930	5.9
1935	5.7
1940	7.8
1945	10.1
1950	12.7
1955	17.3
1960	19.9
1965	23.2
1970	29.5
1975	32.7
1980	34.2
1985	30.9
1990	33.6
1995	34.4
2000	38.3
2005	40.4
2010	37.1

Source: "History of Energy Consumption in the United States, 1775–2009." U.S. Energy Information Administration. http://www.eia.gov/todayinenergy/detail.cfm?id=10. Accessed on March 18, 2014.

Table 1.3 Price of Crude Oil, World Markets, 1920–2008

Year	Price (in 2009 dollars)
1920	32.86
1925	20.55
1930	15.28
1935	15.15
1940	15.59
1945	12.51
1950	15.23
1955	15.45
1960	13.75
1965	12.23
1970	9.94
1975	45.98
1980	95.89
1985	54.95
1990	38.94
1995	23.95
2000	35.50
2005	59.89
2009	61.67

Source: "Historical Crude Oil Prices, 1861 to Present." ChartsBin. http://chartsbin.com/view/oau, based on data from *BP Statistical Review of World Energy, June 2010.* http://www.bp.com/liveassets/bp_internet/globalbp/globalbp_uk_english/reports_and_publications/statistical_energy_review_2008/STAGING/local_assets/2010_downloads/statistical_review_of_world_energy_full_report_2010.pdf. Accessed on March 18, 2014.

Development of the Wind Turbine

The direction of wind-powered technology began to change in the 1930s. Instead of focusing on windmills designed to grind grain, move water, and perform other agricultural and industrial tasks, inventors began to focus on the development of wind turbines, machines for the conversion of mechanical energy to electrical energy. The earliest of those wind turbines (e.g., Blyth and Brush) looked very similar to traditional windmills, even though they had very different functions. But most of the new wind turbines, such as those invented by the Jacobs brothers, bore very little resemblance to pole, tower, and other windmills, or even to the early Blyth and Brush wind turbines.

Research on the use of wind power to generate electricity was strongly influenced by theoretical developments in the young field of aeronautics. As engineers and inventors found ways of making more efficient flying machines, some also began to look for ways in which those developments could be used for the design of more efficient wind turbines. Perhaps the earliest pioneer in this effort was Poul la Cour, from Denmark. Although trained as a meteorologist, La Cour was interested in a wide range of technical and educational problems. Beginning in 1891, he worked on the development of a wind machine that could be used to produce electricity in rural settings, a topic on which he worked until his death in 1908. La Cour's efforts were motivated by the fact that not only most rural areas in Denmark lacked access to electricity (as was the case everywhere in the world) but also that wind power was abundant in Denmark.

La Cour's first inventions looked very much like traditional windmills with, generally, four or six blades mounted on a support tower. (For photographs of La Cour wind turbines, see http://www.poullacour.dk/index-uk.htm.) As his designs improved, he was eventually able to construct wind turbines capable of producing up to 25 kilowatts of power for use on farms and in small villages. These wind turbines became very popular, and by 1900, more than 2,500 had been installed throughout the country, accounting for about a quarter of all the electricity produced in Denmark (Asmus 2001, 42; Poul la Cour 2014). As was to be the case in most other parts of the world, the ready availability of gasoline and other crude oil products soon made wind power economically unsuitable in Denmark, at least until near the end of the twentieth century.

One of the early debates about wind turbines focused on the number of blades best suited for a wind turbine. At first, La Cour flirted with the idea of a two-bladed turbine, as did a number of his contemporaries and successors. For example, inventor brothers John and Gerhard Albers, of Cherokee, Iowa, constructed a wind turbine consisting of two blades in 1927, a device they called the Wincharger. The turbine was designed to charge a six-volt battery and was advertised with the promise that purchasers could "stop spending money on dry batteries!" The inexpensive Albers turbines (one model sold for $15 per unit) soon became widely popular in rural areas of the Midwest. One historian has called them the "Chevrolet of wind turbines" in comparison to the "Jacobs Cadillac" (Gipe 2014; "The Wincharger Corporation" 2014).

La Cour, the Albers, and other early inventors soon discovered that the two-bladed turbine was aerodynamically unstable. Strong winds tended to disrupt the position of the blades on the tower, a problem that was solved by the use of three blades rather than two. In fact, the vast majority of wind turbines now in operation are of the three-blade design. (For a detailed discussion of this issue, see Milborrow 2014.)

Some researchers explored a fundamentally different design for wind turbines, one that drew on the historic vertical-axis plan that had largely been ignored in most twentieth-century inventions. One such design was developed by the Finnish engineer Sigurd Johannes Savonius in the early 1920s. The Savonius design was simple, consisting of two half-metal cylinders, set off from each other on a common axis. (For diagrams of this design, see "Savonius Windmill" 2014.) The two cylinders trapped wind blowing at right angles to the device, causing them and the axis on which they were mounted to turn. The Savonius wind turbine has proved to be especially effective for generating small electrical currents in light winds and continues to be very popular for small-scale applications even today. For example, it is often used as a simple and inexpensive source of electricity in areas where high technology is not available. Savonius devices are also used commonly in ventilation systems for buses, vans, and other heavy vehicles. (See, for example, Abraham et al. 2012; "Flettner Ventilator" 2014). Some highly modified versions of the Savonius design are used in today's world as wind turbines. (See, for example, http://www.archiexpo.com/prod/helixwind/small-vertical-axis-wind-turbines-helical-savonius-rotor-62253-159877.html and http://www.ecosources.info/en/topics/Savonius_vertical_axis_wind_turbine.)

Another type of a vertical-axis wind turbine was designed in the early 1930s by French aeronautical engineer Georges Jean Marie Darrieus. This design consisted of three hoop-shaped metal bands attached to a central vertical post. The force of the wind caused the bands to rotate around the central shaft, at the bottom of which the mechanical energy was converted to electrical energy. Darrieus applied for a U.S. patent for his invention in 1926, a request that was approved on December 8, 1931. (See the patent and the design of the turbine at http://www.freepatentsonline.com/1835018.pdf.) The Darrieus wind turbines used today usually consist of two or three bands and are popularly known as *eggbeaters* because of their resemblance to the kitchen appliance.

Darrieus turbines have important advantages. First the device is built very close to the ground, so its parts are available for regular maintenance at little cost and with almost no inconvenience. Second, the turbine turns with wind with essentially no regard for the direction in which it comes without the risk of equipment disruption experienced with the Savonius device. The Darrieus machine also has a number of disadvantages, however. For example, since it rests close to the ground, it tends to receive the least robust winds blowing at any given time. (In general, the higher the altitude above the ground, the stronger winds tend to blow.) Also, the Darrieus machine will not start operating on its own but requires an impetus from a human or some other source. Other technical problems involved in the construction and maintenance of a Darrieus turbine also somewhat reduce its attractiveness as a power source ("Wind Power from the Darrieus Wind Turbine" 2014).

Darrieus discontinued his research on wind turbines after receiving his U.S. patent, and his invention was largely ignored for the next three decades. In the mid-1960s, however, researchers at the National Research Council of Canada created a wind turbine design that they thought was new and attempted to patent their invention. When their patent application was denied, they found that their "new" device was essentially the same as the one Darrieus had invented thirty years earlier ("A Brief History of Wind Power Development in Canada 1960s–1990s" 2014).

The rediscovered Darrieus machine soon became popular throughout Canada, after which it was imported and tested by the Sandia National Laboratories in Albuquerque, New Mexico. It was eventually adopted for use in some of the large wind power facilities developed in California in the 1970s.

Wind Power Technology: Neglect and Resurgence

The potential of wind power for solving the world's energy problems held the attention of a handful of researchers from the 1930s through the 1960s. But in a practical sense, the technology had less and less of a practical impact on energy production and consumption with each passing year. The reason for this situation was that fossil fuels had become the overwhelmingly dominant source of energy in the United States and other developed nations. None of the alternative forms of energy so near and dear to the hearts of environmentalists—solar, wind, water, and geothermal, for example—had yet been developed sufficiently to compete in the marketplace with coal, oil, and natural gas. None of the vaunted developments in wind technology described earlier could compete on a commercial scale with the internal combustion engine or other fossil fuel devices (see Tables 1.2 and 1.3).

For example, the cost of generating electricity in conventional coal- and oil-powered plants in 1980 was about five cents per kilowatt-hour; for wind-generated electricity, the cost was about 55 cents per kilowatt-hour (McVeigh et al. 1999, Figure 2, page 10; Figure 12, page 22). Under such circumstances, no alternative technology for generating energy in the United States or the rest of the developed world was taken seriously by policy makers, legislators, and others who controlled national economies, all of which was not to deny that a handful of dreamers and optimists saw that the era of fossil fuel could not last forever. A very early example of this position was French engineer Louis Constantin, who wrote in a 1924 essay that:

The Earth's reserves of fuels, solid and liquid, are rapidly being exhausted, and whatever be the hopes, splendid but distant, that give rise to the radioactivity of matter [i.e., nuclear power], this suspended threat to our economic life merits the attention of every thinking man. (Shepherd 1990, 37, quoting Constantin 1924)

Constantin saw wind power as a possible way of providing energy in the future, and he devoted much of his life to studying the technology needed to make this hope a reality. Although a resident of France, he contributed a number of papers and obtained a number of patents in the United States, on occasion working in connection with the National Advisory Committee on Aeronautics, an agency founded in 1915 to sponsor and promote aeronautical research in the United States (see, for example, Constantin 1921; "Louis Constantin" 2014).

Warnings about the limited availability of fossil fuels came from a number of energy experts and environmentalists also. For example, American geophysicist M. King Hubbert began writing about the approach of "peak oil" in the 1950s. The term *peak oil* refers to a period of time at which a nation (or the world) has produced the maximum amount of petroleum it can from its natural reserves. Beyond that point, oil production decreases until it reaches zero for all practical purposes. Hubbert's concept of peak oil implied that nations and the world would eventually have to begin thinking about alternative sources of energy, such as solar, wind, and geothermal, and the sooner political leaders and researchers began to do so, the better able nations and the world would be able to deal with the inevitable collapse of the age of fossil fuels. (Hubbert's original speech on peak oil can be found at Hubbert 2014; see also Deffeyes 2009.)

In fact, it took something far more threatening and more immediate that the hopeful research of men and women like Constantin or the dire warnings of environmental prophets of gloom and doom to actually awaken the world's political systems to the potential alternative energy systems, such as wind power. And that something occurred in 1973 with the Arab oil embargo.

The 1973 Arab oil embargo was an action taken by members of the Organization of Arab Petroleum Exporting Countries (OAPEC) against Canada, Japan, the Netherlands, the United Kingdom, and the United States in retaliation for the support from these countries for the nation of Israel in its war against Egypt, Syria, and other Arab nations—the Yom Kippur War—that lasted from October 6 to 25, 1973 (Alhajji 2005). As a major supplier of crude oil to the industrialized world, OAPEC decided that it could use the cost of oil as a tool in its battle against Israel while, at the same time, bolstering the economies of its member nations. In the nations affected by the embargo, the price of crude oil skyrocketed at a rate never before seen in history (see Table 1.3). In the United States, for example, the price of crude oil rose from $1.21 per barrel in 1970 to $11 per barrel at the end of 1974 (Adelman 2014, Abstract). The action by OAPEC acted as a wake-up call to developed nations, as perhaps nothing else had, or could have done, of their dependence on fossil fuels. Whatever other effects the embargo may have had, it clearly showed the necessity for at least beginning to explore the options provided by alternative sources of energy, such as solar, wind, and geothermal.

The 1973 oil embargo had a host of short-term and long-term political, social, economic, environmental, and other effects in the United States and other nations affected by the embargo. For example, the U.S. Congress passed the Emergency Highway Energy Conservation Act of 1974, which, among other things, reduced the speed limit on federal highways to 55 miles per hour. The Congress also passed a comprehensive National Energy Act in 1978 dealing with a variety of energy-related issues. The Energy Policy and Conservation Act of 1975 also established the Strategic Petroleum Reserve, a fuel storage area located at four sites in the southern United States designed to provide an emergency resource in case of future fuel embargoes ("Petroleum Chronology of Events" 1970–2000).

Governmental Support for Wind Energy

Within a matter of months after the announcement of an oil embargo by OAPEC, President Richard M. Nixon began to assemble his response to this threat to the U.S. economy. On November 7, 1973, he announced plans for his Project Independence, a federal program that would rival the Manhattan Project (in which the world's first nuclear weapons were designed and built) in its impact on American capabilities in the field of energy. He saw a number of elements as necessary to make the United States independent of foreign energy supplies (e.g., oil from the Arab states), including reduced speed limits and home use of electricity, greater conservation efforts by industry and individuals, and an increase in the search for and collection of fossil fuels from American territories. He also mentioned the need for greater attention to the contribution that alternative energy sources could make to future American energy independence. In his 1980 retrospective book, *The Real War,* Nixon explained that the goal of his Project Independence was "to stimulate the production of energy from renewable sources such as nuclear power, and, in the short run, to cut back our dependency on unreliable foreign suppliers of oil." He further noted that he was "the first president to propose a wide-ranging energy program" (Nixon 1980, 222–223).

(The terms *alternative* and *renewable* are sometimes used interchangeably, although they do not strictly mean the same thing. The broader term *alternative* generally refers to any type of energy source other than the fossil fuels of coal, oil, and natural gas. By contrast, the term *renewable* generally refers to energy sources obtained from natural processes that will presumably always be available to humans, such as solar, wind, tidal, geothermal, and most forms of biomass.)

As the character of Project Independence became clearer, it was obvious that the role of renewable or alternative energy was, to be generous, minimal. A 119-page summary of the program, developed by the Federal Energy Administration in 1974, for example, mentioned three ways of dealing with the energy crisis: increasing domestic energy supplies, extending programs for conservation of energy resources, and developing "emergency programs" such as the Strategic Petroleum Reserve. It mentioned increased attention to renewable energy sources in general only once and referred to solar power a total of five times; geothermal power, seven times; and nuclear power, thirty-seven times. Wind power was not mentioned in the report at all (*Project Independence: A Summary* 1974). The rationale for the short shrift given to alternative energy sources was that such resources would not reach a stage of development at which they could contribute to the nation's energy needs until after 1985, more than a decade into the future (*Project Independence: A Summary* 1974, 6).

Nixon's (at least verbal) commitment to the importance of renewable energy sources was to become an ongoing mantra for almost every U.S. president who followed him, usually with about as little results as Nixon had experienced. In his first State of the Union speech, Nixon's successor, President Gerald Ford, echoed his predecessor's call for a new energy policy for the United States in which alternative energy was to play a role, although, as the speech went on, it was a minor and undefined role. Still, Ford made enough of an effort to promote alternative energy sources that, when he died in 2007, *New York Times* columnist and Pulitzer Price–winning economist Thomas Friedman was inspired to call him "the first energy president" (Friedman 2007).

One of the most important accomplishments in the promotion of alternative energy sources during the Ford administration was the creation of the Energy Research and Development Administration. The new agency was to take over a number of the responsibilities of the former Atomic Energy Commission in the promotion of a variety of new energy sources. Among those energy sources mentioned by Ford in his executive order activating the agency were "fossil [fuels], nuclear fission and fusion, solar, and geothermal." Again, no specific mention of wind energy was made (Buck 2014; Ford 2014)!

Another important development during the Ford administration was the passage of the Solar Research, Development, and Demonstration Act of 1974, which authorized the establishment of the Solar Energy Research Institute (SERI) "for the purpose of resolving the major technical problems inhibiting commercial utilization of solar energy in the United States" (Public Law 93–473). In 1991, President George H. W. Bush redesignated SERI as the National Renewable Energy Laboratory (NREL) and broadened its mandate to include research on a greater variety of alternative energy sources, including wind energy ("The History of Solar" 2014).

Yet another important step forward in the development of alternative energy during the Ford administration was the origin of a program on wind energy under the auspices of the National Aeronautics and Space Administration (NASA). That program, also motivated by the 1973 Arab oil embargo, was designed to test the design of a number of wind turbine models for the possible use of large-scale generation of electrical energy. The research was carried out primarily at the NASA Lewis Research Center (now the Glenn Research Center) in Sandusky, Ohio. During the life of this program, from 1974 to 1992, NASA spent about $330 million on turbine research (Douthwaite 2002, 97). The NASA program tested seven different models of wind turbines, all two-bladed designs. One model was never built and tested, one was tested only in prototype, and the other five were field tested (Thomas 1982).

Momentum for an increased emphasis on alternative energy research continued under the administration of President Jimmy Carter, from 1977 to 1981. In one of the earliest acts of his administration, Carter gave a speech on April 18, 1977, in which he declared that efforts to deal with the nation's energy crisis were serious enough to be treated as "the moral equivalent of war," a comment that was to become one of the keynotes of his administration's legislative program. Among the ten principles enunciated by Carter in his speech, one (the last) was that the nation would have to begin an aggressive research program "to develop the new, unconventional sources of energy we will rely on in the next century" (Carter 2014).

Probably the most concrete expression of this energy policy came later in the same year when the U.S. Congress passed and Carter signed the Department of Energy Reorganization Act (Public Law 95–91). That act consolidated the responsibilities and activities of more than thirty different federal agencies into a single, new cabinet-level department. Carter's strategy was to emphasize the importance of energy policy by creating a massive new agency with authority to administer all aspects of the nation's energy activities. Mention of alternative and renewable energy sources was minimal in the organization act, but those few mentions were adequate to establish a legislative basis for a great variety of future research activities on alternative energy sources. (Nuclear energy was mentioned twenty-three times in the act; solar power and geothermal power, four times each; and wind power, not at all [Public Law 95–91].)

The Declining Interest in Alternative Energy

Enthusiasm for the potential role of alternative energy sources in the nation's future energy equation, however, turned out to be short-lived. In a relatively short period of time, Carter's successor, Ronald Reagan, largely disassembled the federal effort to promote research on alternative energy. Reagan made his position on alternative energy clear in a presidential debate with then-president Jimmy Carter on October 28, 1980. He expressed the view that research on alternative energy programs in the preceding decade had been a waste of money. "It hasn't produced," he said, "a quart of oil or a lump of coal, or anything else in the line of energy" ("October 28 1980, Debate Transcript" 2014).

Once elected, Reagan was true to this view of alternative energy. In a relatively short period of time, he cut the Department of Energy's budget for alternative energy and conservation by half, reduced the budget of NREL by nearly 90 percent, eliminated the wind investment tax credit (more about this item in the next chapter), reduced funding for research on solar photovoltaic cells by two-thirds, and removed the solar panels on the White House (Parry 2014). The last of these actions was almost certainly symbolic only. President Carter had thirty-two solar panels installed on the top of the White House on June 20, 1979, in order to harvest the energy of the sun for local applications. That action had almost no effect at all on the White House energy budget, but it did symbolize Carter's commitment to the development of alternative energy sources. Seven years later, essentially without comment, Reagan ordered the panels to be removed—again, an act apparently designed for symbolic, rather than practical, purposes (Biello 2010).

The concrete result of Reagan's view on renewable energy (and that of his successor, President George H.W. Bush) can be seen in budget allocations for this line of research over the next twelve years. The average annual appropriation for research on renewable energy during the Carter administration (FY 1978–FY 1981) was $1.290 billion. During the eight-year Reagan administration (FY 1982–FY 1990), the average annual appropriation fell to $253 million, and under Bush (FY 1991–FY 1995), it fell slightly again to $209 million annually ("R&D Priorities within the Department of Energy" 2014, Table IV.2.2).

Under the presidential administrations of George W. Bush and Barack Obama, there has been once more an increasing emphasis on alternative energy research. The federal budget for this item essentially doubled between FY 2007 and FY 2010, with an increase in funding from $717 million in the former year to $1.409 billion in the latter year. Table 1.4 shows the allocation of this funding among various types of alternative energy sources

State Actions on Wind Energy

In the absence of strong federal policies and financial support for the development of wind power in the United States, a number of states have developed a variety of incentives for

Table 1.4 Funding for Renewable Energy Sources, FY 2007 to FY 2010 (in millions of dollars)

Program	FY 2007	FY 2010
Wind	58	166
Solar	171	348
Hydrogen Technology	211	205
Biofuels and Biomass	268	537
Geothermal	9	100
Hydroelectric	0	52
Total	717	1,409

Source: Direct Federal Financial Interventions and Subsidies in Energy in Fiscal Year 2010. U.S. Energy Information Administration, July 2011, Table 13. http://www.eia.gov/analysis/requests/subsidy/pdf/subsidy.pdf. Accessed on March 22, 2014.

encouraging wind research and development within their own borders. These incentives include clean energy funds; reduction in or exemption from state sales taxes; investment tax credits; property tax reductions; accelerated depreciation; direct production incentives; direct investment incentives (i.e., grants); government-subsidized loans; so-called standard offer contracts that guarantee prices into the future; net metering or net billing; site prospecting, review, and permitting programs that reduce the siting challenges in building new wind facilities; renewable set-asides; surcharges on electric bills to pay for the development of renewable energy facilities; green marketing, in which consumers agree to pay a higher price for "environmentally friendly" electricity; and state mandates, which require power companies to produce some fraction of their energy from renewable sources over some period of time.

One of the most popular forms of incentives adopted by states is the Renewable Portfolio Standard (RPS). An RPS is a formally adopted state policy that requires an energy producer to include some fraction of its annual electrical output from renewable energy sources. As of mid-2014, thirty-five states, the District of Columbia, and Puerto Rico had adopted some form of RPS. An example of an RPS is that of Wisconsin's, which originally (1998) required investor-owned utilities and electric cooperatives to obtain 2.2 percent of the electricity sold to customers to come from renewable energy resources by 2012. That standard was changed in 2006 to require 10 percent of all electricity sold to come from renewable resources by 2015. (The master source of information about state incentives for research and development on renewable energy is the Database of State Incentives for Renewables and Efficiency [DSIRE] at http://www.dsireusa.org/. For information on state-specific RPS programs, see http://www.dsireusa.org/documents/summarymaps/RPS_map.pdf. For information on the Wisconsin program described here, see http://www.dsireusa.org/incentives/incentive.cfm?Incentive_Code=WI05R&re=1&ee=1.)

RPSs may be either mandatory or voluntary. As the terms suggest, electricity suppliers may either be required to abide by a state's RPS or choose to do so or not. States with mandatory policies routinely adopt some sort of penalty for utilities that do not meet state standards. As an example, the state of Oregon's RPS provides for the assessment of a fine called the Alternative Compliance Payment (ACP) for utilities not in compliance with the state's RPS. The amount of the ACP fine is determined on a regular basis by the state's Department of Energy ("Summary of Oregon's Renewable Portfolio Standard" 2014).

Wind Turbine Technology

At one point in history, the design and construction of wind machines was primarily a matter of trial and error. Inventors built windmills with different numbers of blades, sails made of different materials, blades and sails of various design, axis oriented in varying directions and with differing physical characteristics, and so on. The machines that worked best were then used as models for later developments in windmill design.

With the advance of wind turbines, however, the process of design and invention changed. Engineers and inventors began to draw on the growing body of scientific knowledge about the properties of wind, characteristics of various building materials, the generation of electricity from kinetic energy, the movement of connecting parts of a wind machine, and other factors involved in the construction of a wind turbine. Today, the design and construction of wind turbines is a highly sophisticated, scientific and technological process, with engineers employing the latest information from physics, meteorology, chemistry, and other fields of science.

The governing principle that underlies all wind turbine design is called Betz's law, named after German physicist Albert Betz. In 1919, Betz derived an equation that expresses the amount of power that can be generated by a wind turbine, given certain basic factors. That equation is now known as the *wind power equation,* and may take a variety of forms. Perhaps the simplest statement of the equation is:

$$P = 1/2\rho AV^3$$

where P = the power generated by the turbine, ρ = the density of air surrounding the turbine, A = the area swept out by the turbine blades, and V = the velocity of the wind. This equation can be adapted to include other factors involved in the operation of the turbine, such as the mechanical efficiency of the working parts of the turbine, such as the gear box and the generator.

This equation is so fundamental to research on wind turbines that it is possible to find automated programs that quickly calculate the value of P, given known values for the other variables. Such a program can be found, for example, at the Web page of designer Jimmy Raymond (http://www.ajdesigner.com/phpwindpower/wind_generator_power.php#ajscroll). Try substituting the following arbitrary values in the wind equation in the spaces provided on the Web site:

$\rho = 1.225$; $A = 200$; $V = 30$; $Ng = 0.5$; $Nb = 0.5$; $Cp = 0.5$,

Unless you are using this format throughout, in which case, keep as set and notice that the wind power obtained from this set of variables is 413,437.5 watt. (Other Web sites with specialized wind power calculators include "Wind Energy: Harnessing the Air in Motion," http://www.windpowercalculator.mywindpowersystem.com/; Windpower Program, http://www.wind-power-program.com/download.htm; and the Swiss Wind Power Data Web site, http://wind-data.ch/tools/.)

A crucial additional factor in this equation not yet mentioned is Cp, sometimes known as the *coefficient of performance.* In addition to deriving the wind power equation, Betz made an important discovery about the practical applications of his law. He found that no wind turbine is capable of capturing all of the wind energy that strikes it. Such a discovery should hardly be surprising, since very few (if any) natural events occur with

100 percent efficiency. Some energy is always converted to other forms ("lost") during such events. In 1920, Betz was able to calculate that the maximum amount of energy *any* wind turbine can capture is 59.3 percent (0.593) of the wind energy that strikes the turbine's rotors. So the original, simplest wind energy shown earlier is only the expression of ideal conditions, which are never present in the real world. A more realistic version of that equation for real wind turbines in everyday life must also take into consideration how efficiently that particular wind turbine is available to convert wind energy into mechanical energy in the turbine, or:

$P = 1/2\rho C_p A V^3$.

One additional point to note from the Raymond Web site mentioned earlier is the additional benefits of having the answer expressed in a variety of units. The reason for this additional section is that energy and power can be expressed in a variety of units in either the metric (SI) or English system. The SI (Le Système international d'unités, or International System of Units) is used by every country in the world with a handful of exceptions (including the United States) and by all scientists worldwide. The English system of measurement (also known as the British or customary system) is used only in the United States, Liberia, and Myanmar (the former Burma). It is often used by engineers and technicians. The units used for energy, defined as the ability to do work, and for power, defined as the amount of energy consumed or produced per unit of time, are shown in Table 1.5.

In practice, a variety of other units may be used, especially in the SI system, primarily because the units shown in Table 1.5 represent very small quantities of energy and power (such as those expended by an ant pushing on a grain of rice). Some examples of these more common units for power are kilowatt (103, or a thousand watts), megawatt (106, or a million watts), gigawatt (109, or a billion watts), terawatt (1012, or a trillion watts), and petawatt (1015, or a quadrillion watts).

Table 1.5 Units of Energy and Power

Unit	Symbol	Other Relationship(s)	Symbol
Energy			
Joule	J	kilogram × meter^2 ÷ second^2	kgm^{2}s^{-2}
Kilowatt-hour	kWh	kW × h 1 kWh = 3.6 × 10^{6}J	
Foot-pound	ft-lb	force that moves a mass of one pound a distance of one foot	
Power			
Watt	W	kilogram × meter^2 ÷ second^3	kgm^{2}s^{-3} Js^{-1}
Foot-pound-second	ft-lb-s^{-1}	force that moves a mass of one pound a distance of one foot in one second	
Horsepower	hp	1 hp = 550 ft-lb-s^{-1} = 746 W	

Components of a Wind Turbine

As noted previously, the structure of a wind turbine is simple in concept. It consists of an airplane-engine-type structure called the *nacelle* mounted on a tall tower. Attached to the front of the nacelle is a *rotor,* to which are attached usually two or three *blades.* (A number of excellent diagrams of wind turbines are available on the Web, perhaps best being those provided by the Office of Energy Efficiency and Renewable Energy at http://energy.gov/eere/wind/how-do-wind-turbines-work.) When wind strikes the blades, it causes them to rotate on the axis of the rotor, a motion that is transmitted to a pair of shafts running down the center of the nacelle. The relatively low-speed rotation induced in the first shaft is speeded up in a gear box that connects the slow-speed shaft to a high-speed shaft. The high-speed shaft, in turn, is connected to an electrical generator, where the kinetic motion of the wind is ultimately converted to electrical energy. The electrical energy is then carried away from the turbine by a system of wires, through which it is transmitted to consumers.

The nacelle also contains mechanisms for controlling potentially destructive damage to the turbine caused by high wind speeds. This damage can result from two types of undesirable motions within the turbine—pitch and yaw. Pitch is the tendency of the nacelle to experience up-and-down motions along the axis from the front of the nacelle to the rear. Yaw is the tendency of the nacelle to experience a horizontal motion perpendicular to the axis of the nacelle. Either of these motions can become violent enough to cause physical damage to the nacelle, so that auxiliary devices are needed to counteract and moderate each effect.

Other auxiliary equipment on the wind turbine include devices that monitor wind speed and control operation of the turbine at various wind speeds. The anemometer and wind vane attached to the rear of the nacelle, for example, allow for the measurement of the speed and direction of the wind, so that it can constantly be oriented into the wind. The controller is a device for turning the turbine on and off, the latter usually at speeds greater than about 55 miles per hour. In winds stronger than that speed, physical damage to the turbine is a possibility, so the device is simply shut down in strong winds.

Wind Farms

In their heyday, windmills were used almost exclusively for the power needs of individual homes, farms, or businesses. They were installed to grind grain, lift water, or produce electricity for a handful of individuals. As enthusiasm for wind power grew after the 1973 oil embargo; however, researchers, engineers, and inventors began to think more ambitiously about the uses to which wind power could be put. They imagined wind systems that were large enough to supply electrical energy to whole communities or to large industrial facilities. At the same time, the (what turned out to be short-lived) enthusiasm of the U.S. government for the support of wind projects provided the financial support needed to build and test such systems. The fundamental concept was that, while a single wind device could supply only a small amount of electricity at a time, some combination of a dozen or more machines could produce a much larger supply. The first such system, called a *wind farm,* was installed in December 1980 at Crotched Mountain, New Hampshire. The system was developed and operated by U.S. Windpower (later Kenetech) and consisted of twenty wind turbines, each with a rated energy output of 30 kilowatts. The project lasted only about a year, however, as the developer had misjudged wind conditions in the area, and turbine equipment experienced too many breakdowns to justify continuation of the project (Mooiman 2014; "New England Wind Forum" 2014; an interesting photograph of the original wind farm turbines is available at http://www.nashuatelegraph.com/granitegeek/1003657-468/a-great-picture-of-the-worlds-first.html).

The U.S. Department of Energy has summarized the growth of wind farms in the United States on an interactive Web page at http://energy.gov/articles/wind-farm-growth-through-years. That site indicates that the first utility-scale wind farm in the United States was established in Southern California in 1975. It consisted of a single turbine capable of supplying electrical energy to 4,149 homes. Growth of wind farms in the United States over the succeeding four decades can be tracked through this interactive program, reflecting the data shown in Table 1.6.

It should be noted that all of the first wind farms were built in California. The first wind farm constructed in another state was not installed until 1994, when the Buffalo Ridge Wind Farm near Lake Benton went into operation. At that point, fifty-two wind farms in California were already in operation (Shaffer 2014).

Table 1.6 Growth in Wind Farms in the United States, 1975–2012

Year	Wind Farms	Homes Supplied
1975	1	4,149
1981	2	8,575
1982	4	13,500
1983	8	109,000
1984	15	146,000
1985	23	175,000
1986	30	196,000
1987	36	225,000
1988	37	230,000
1989	42	257,000
1990	46	294,000
1991	47	314,000
1992	48	315,000
1993	49	317,000
1994	53	342,000
1995	54	349,000
1996	55	357,000
1997	58	361,000
1998	66	405,000
1999	91	579,000
2000	97	592,000
2001	135	948,000
2002	149	1,100,000
2003	185	1,600,000
2004	202	1,700,000
2005	226	2,200,000
2006	270	2,900,000
2007	318	4,200,000
2008	416	6,500,000
2009	523	8,900,000
2010	581	10,000,000
2011	677	12,000,000
2012	815	15,000,000

Source: Wind Farm Growth through the Years. U.S. Department of Energy. http://energy.gov/articles/wind farm growth through years#buttn. Accessed on March 23, 2014.

The International Scene

Denmark

The story of the development of wind power in the United States is not one that can be told for all, or even some, other countries of the world. Indeed, the wind power industry has had a somewhat different history in various countries. The nation perhaps most commonly mentioned as a leader in the development of wind power facilities is Denmark. Such is the case to a large extent because of the geographical setting of the country, along with its natural resources, or lack of them. Denmark has a coastline of 7,314 kilometers for a total land area of 42,434 square kilometers, or one kilometer of coastline for just under every six square kilometers of land area. By comparison, the United States has 1 kilometer of coastline for every 460 square kilometers of land area and China, 1 kilometer of coastline for every 630 square kilometers of land area (*The World Factbook* 2014).

Since the wind tends to blow more regularly and with greater force along the coast, Denmark would appear to be a logical place for the development of wind energy facilities. At the same time, the country has virtually no other natural energy resources of its own: no coal, oil, or natural gas reserves; no rivers or geothermal resources; and no natural sources of uranium for nuclear power plants.

It is hardly surprising, then, that imaginative inventors like Poul la Cour decided to take advantage of Denmark's one plentiful source of energy—the wind. For a short period of time, as noted earlier, La Cour's program to bring wind-generated electricity to homes, farms, and small communities was very successful. Before long, however, the program was overwhelmed by the development of coal-fired, electricity-generating power plants. After World War I, La Cour's 250 wind plants began to decrease in number, to 75 in 1920 and then to 25 in 1940 (Vestergaard, Brandstrup, and Goddard 2004).

World War II once more brought fossil fuel shortages to Denmark, and interest in wind-generated power once more increased. This interest led to the development of a host of new wind technologies that made Denmark the world's leader in the field. In a now-familiar pattern, however, the end of the war and the abundance of fossil fuels that followed once again caused wind generation to fall into disfavor until, as in the United States, the Arab oil embargo dramatically increased the price of oil. This time, the emphasis on wind energy that developed did not dissipate but has continued to the present day. Beginning at the end of the twentieth century, wind-generated electricity has played an increasingly important role in the Danish economy, until today; it now accounts for a third of all the electrical energy used in the nation.

Table 1.7 Wind Energy Production in Denmark, 2003–2013

Year	Wind Energy as Share of Total Electricity Production (percent)	Accumulated Wind Capacity (MW)		
		Onshore	Offshore	Total
2003	15.97	423	2,460	2,883
2004	18.76	423	2,462	2,885
2005	18.66	423	2,484	2,907
2006	16.97	423	2,496	2,919
2007	19.86	423	2,498	2,921
2008	19.33	423	2,567	2,990
2009	19.40	661	2,675	3,335
2010	22.00	868	2,831	3,699
2011	28.27	871	3,020	3,891
2012	30.08	922	3,194	4,116
2013	33.20	1,271	3,501	4,772
2014		*1,271*	*3,754*	*5,025*
2015		*1,271*	*3,954*	*5,225*

Numbers in italics are estimates.

Source: "The Danish Market." Danish Wind Industry Association. http://www.windpower.org/en/knowledge/statistics/the_danish_market.html. Accessed on March 25, 2014. Data used by permission of the Danish Wind Industry Association.

Africa

In many nations around the world, the story of wind power development is very much different from that in Denmark. Such is the case for nearly all developing countries, for fairly obvious reasons. Most such nations struggle to provide their populations with the basic necessities of food, water, medical care, housing, and other essentials of daily life. They seldom have the technical capabilities, financial resources, personnel, and, sometimes, the will to be concerned with the development of alternative energy sources.

In Africa, for example, two of the nations with the greatest potential for wind-generated electricity are Mauritania and Somalia. In Mauritania, for example, some studies have shown that wind power could produce four times as much energy as all the oil it currently uses for energy production. In Somalia, wind power appears to have the potential to provide 90 percent of that nation's total energy needs (Mukasa et al. 2014, 2). In both countries, geographic, political, social, economic, and other factors have prevented the development of these wind resources. Other studies have found that wind power has the potential to supply a significant portion of the energy needs in a number of other African nations, especially Angola, Mozambique, Namibia, South Africa, and Tanzania (Mukasa et al. 2014, 2).

One of the few nations that have had any success at all in developing its wind resources is Egypt. Table 1.8 shows the development of wind resources in Egypt from 1997 to 2013.

Table 1.8 Installed Wind Power Capacity in Egypt, 1997–2013 (MW)

Year	Production Capacity (MW)
1997	6
1998	6
1999	36
2000	69
2001	69
2002	69

(Continued)

Table 1.8 *(Continued)*

Year	Production Capacity (MW)
2003	180
2004	145
2005	145
2006	230
2007	310
2008	390
2009	430
2010	550
2011	550
2012	550
2013	550

Source: "Egypt: Production Capacities." The Windpower. http://www.thewindpower.net/country_en_22_egypt.php. Accessed on March 25, 2014. Data based on information from annual World Wind Energy Reports of the World Wind Energy Association. Used by permission of the World Wind Energy Association, www.wwindea.org.

China

Perhaps the most interesting developments in the use of wind power have taken place in China. The Chinese wind energy program began in 1986 with a demonstration program that focused on the development of relatively small wind turbines for local applications. The next step forward in the program involved connecting local wind generators to the main power grid, where excess electricity generated by wind turbines could be sold to electric utilities. In 2001, central planners committed the country to an ambitious wind energy development program called Ride the Wind Programme, with the goal of developing a national wind capacity of 1,000 MW by 2001 ("China: Market Overview" 2014).

The program was not successful in achieving that goal, as indicated in Table 1.9, attaining a capacity in 2001 of less than half that number, 404 MW. The government continued to emphasize the role of wind power in China's future energy equation, however, and a number of economic incentives were developed to improve wind turbine technology, encourage turbine construction, and make wind energy available for a broader array of applications. A key turning point came in 2005 when the government adopted the first Renewable Energy Law, which laid out future expectations of the role of renewable energy in the national economy and established a number of economic policies to ensure that these goals would be met. Within a matter of years, the effects of the new policies became obvious, with growth in wind capacity rising from 2,559 MW in 2006 to 62,733 MW in 2011 ("China: Market Overview" 2014). By 2010, China had become the world's leading producer of wind energy, with a rated capacity of 44,733 MW. At that point, the nation had been doubling its wind energy output every year since 2006 and was expected to continue to hold its position as world leader in wind energy output until at least 2020 (Bayar 2014).

Table 1.9 Growth in Wind Power Capacity in China, 2001–2012 (MW)

Year	Installed Capacity (MW)
2001	404
2002	470
2003	568
2004	765
2005	1,272
2006	2,559
2007	5,910
2008	12,020
2009	25,805
2010	44,733
2011	62,364
2012	75,324

Source: Global Wind Report: Annual Market Update 2012. Global Wind Energy Council, 31. Available online at http://www.gwec.net/wp–content/up loads/2012/06/Annual_report_2012_LowRes.pdf. Accessed on March 25, 2014. Used by permission of the Global Wind Energy Council.

Conclusion

The status of wind power in the United States and other parts of the world had changed dramatically in the period from the early twentieth century to the first decades of the twenty-first century. Wind devices had evolved from relatively simple, inexpensive windmills essential, but largely limited to, small farms and businesses to giant wind farms that covered dozens or hundreds of acres serving thousands of households and industrial operations. But this growth has not come without a number of concomitant problems, such as finding ways to deliver wind power at prices comparable to those from other sources; avoiding unacceptable environmental damage, such as the destruction of wildlife and the despoliation of the land; and possible health effects on humans. Chapter 2 will review some of these problems as they affect the development of wind power, not only in the United States but also in developed and developing nations throughout the world.

References

Abraham, J. P., et al. 2012. "Summary of Savonius Wind Turbine Development and Future Applications for Small-Scale Power Generation." *Journal of Renewable and Sustainable Energy.* 4(4): http://scitation.aip.org/content/aip/journal/jrse/4/4/10.1063/1.4747822. Accessed on March 19, 2014.

Adelman, M. A. "The First Oil Price Explosion, 1971–1974." http://dspace.mit.edu/bitstream/handle/1721.1/50146/28596081.pdf?sequence=1. Accessed on March 20, 2014.

Alhajji, A. E. 2005. "The Oil Weapon: Past, Present, and Future." *Oil & Gas Journal.* 103(17): 22–25.

al-Hasan, Ahmad Yousef, and Donald Routledge Hill. 1986. *Islamic Technology: An Illustrated History.* Cambridge, UK; New York: Cambridge University Press.

Ambler, John, and John Langdon. 1994. "Lordship and Peasant Consumerism in the Milling Industry of Early Fourteenth-Century England." *Past & Present.* 145(1): 3–46.

"America's Cup: History." America's Cup. http://www.americascup.com/en/history.html. Accessed on March 14, 2014.

Asmus, Peter. 2001. *Reaping the Wind: How Mechanical Wizards, Visionaries, and Profiteers Helped Shape Our Energy Future.* Washington, DC: Island Press.

Baker, T. Lindsay. 1985. *A Field Guide to American Windmills.* Norman: University of Oklahoma Press.

Baker, T. Lindsay. 2007. *American Windmills: An Album of Historic Photographs.* Norman: University of Oklahoma Press.

Bayar, Tildy. "Despite Slowdown, China to Hold Wind Power Market Leadership to 2020." *RenewableEnergy.World.com.* http://www.renewableenergyworld.com/rea/news/article/2013/08/despite-slowdown-china-to-hold-wind-power-market-leadership-to-2020. Accessed on March 25, 2014.

"Beginnings: First Generation of Design (1920–1931)." Jacobs Wind Electric Co., Inc. http://www.jacobswind.net/history/beginnings. Accessed on March 17, 2014.

Berend, Ivan. 2013. *An Economic History of Nineteenth-Century Europe: Diversity and Industrialization.* Cambridge, UK; New York: Cambridge University Press.

Biello, David. 2010. "Where Did the Carter White House's Solar Panels Go?" *Scientific American.* http://www.scientificamerican.com/article/carter-white-house-solar-panel-array/. Accessed on March 22, 2014.

Breasted, James Henry. 1906. *Ancient Records of Egypt.* Chicago: The University of Chicago Press. http://digitalcase.case.edu:9000/fedora/get/ksl:breanc00/breanc00.pdf. Accessed on March 14, 2014.

"A Brief History of Wind Power Development in Canada 1960s–1990s." http://www.uoguelph.ca/engineering/sites/default/files/resources_History%20Canada.pdf. Accessed on March 19, 2014.

Buck, Alice. "A History of the Energy Research and Development Administration." U.S. Department of Energy. Office of History and Heritage Resources. http://energy.gov/sites/prod/files/ERDA%20History.pdf. Accessed on March 24, 2014.

Carter, Jimmy. "Proposed Energy Policy." http://www.pbs.org/wgbh/americanexperience/features/primary-resources/carter-energy/. Accessed on March 21, 2014.

Carter, W. Hodding. 2000. *A Viking Voyage.* New York: Ballantine.

"Charles F. Brush." Green Energy Ohio. http://www.greenenergyohio.org/page.cfm?pageId=341. Accessed on March 17, 2014.

Chatterton, Edward Keble. 1912. *Fore and Aft: The Story of the Fore & Aft Rig from the Earliest Times to the Present Day.* London: Seeley, Service & Co.

Chatterton, Edward Keble. 2010. *History of Sailing Ships: The Story of Their Development from the Earliest Times until the 19th Century.* Bremen, Germany: Salzwasser-Verlag Gmbh (Reprint of 1909 edition).

"China: Market Overview." International Renewable Energy Agency. https://www.irena.org/DocumentDownloads/Publications/IRENA_GWEC_WindReport_China.pdf. Accessed on March 25, 2014.

Clemmons, Maureen, and Dan Cray. 2001. *Soaring Stones: How a Backyard Scientist and Her Kite-powered Pyramid Theory Took On the World of Egyptology.* Washington, DC: National Geographic Society; London: Hi Marketing.

"Club History." Royal Cork Yacht Club. http://www.royalcork.com/club-history/. Accessed on March 14, 2014.

Constantin, Louis. 1921. "Wind Vane with Various Applications." National Advisory Committee for Aeronautics. http://ntrs.nasa.gov/archive/nasa/casi.ntrs.nasa.gov/19930083142.pdf. Accessed on March 20, 2014.

Constantin, Louis. 1924. "Le Vent." *La Nature.* 52(Part I): 395–400.

Deffeyes, Kenneth S. 2009. *Hubbert's Peak: The Impending World Oil Shortage,* updated edition. Princeton, NJ: Princeton University Press.

Delisle, Leopold. 1851. "On the Origin of Windmills in Normandy and England." *Journal of the British Archaeological Association.* 6: 403–406.

Derry, T. K., and Trevor I. Williams. 1993. *A Short History of Technology from the Earliest Times to A.D. 1900.* New York: Dover Publications. (Reprint of the original 1960 edition.)

Douthwaite, Boru. 2002. *Enabling Innovation: A Practical Guide to Understanding and Fostering Technological Change.* London: Zed Books.

Duell, Mark. "Still Standing after 350 Years: Inside England's Oldest Smock Windmill Saved from Collapse and Painstakingly Restored." *Mail Online.* http://www.dailymail.co.uk/news/article-2398266/How-worlds-oldest-smock-dating-1650-collapse-volunteers-restored-it.html. Accessed on March 16, 2014.

"Energy Time Machine." http://www.energyquest.ca.gov/time_machine/time_machine_hydro1.php?start=25. Accessed on March 15, 2014.

"Explore a Pharaoh's Boat." 2009. Nova. http://www.pbs.org/wgbh/nova/pharaoh/expl-nf.html. Accessed on March 13, 2014.

Flettner, Anton. 1926. *The Story of the Rotor.* New York: F. O. Willhofft.

"Flettner Ventilator." http://www.flettner.co.uk/. Accessed on March 19, 2014.

Forbes, Robert J. 1965. *Studies in Ancient Technology,* vol. 2. Leiden, the Netherlands: Brill.

Ford, Gerald. "Statement Announcing Activation of the Energy Research and Development Administration and the Nuclear Regulatory Commission, January 15, 1975." The American Presidency Project. http://www.presidency.ucsb.edu/ws/index.php?pid=4949. Accessed on March 21, 2014.

Friedman, Thomas L. 2007. "The First Energy President." *New York Times*. January 5, A17.

Gasch, Robert, and Jochen Twele. 2012. "Historical Development of Windmills." In Robert Gasch and Jochen Twele, eds. *Wind Power Plants: Fundamentals, Design, Construction and Operation,* 2nd ed. Berlin; Heidelberg: Springer.

Gipe, Paul. "Wind-Works." http://www.wind-works.org/cms/index.php?id=530. Accessed on March 19, 2014.

Golding, Edward W. 1976. *The Generation of Electricity by Wind Power.* London: E. & F. N. Spon; New York: Halsted Press.

"Herodotus: The Second Book of the Histories, Called Euterpe." http://www.cheops-pyramide.ch/khufu-pyramid/herodotus.html. Accessed on March 13, 2014.

"History." Illinois Windmills. http://www.illinoiswindmills.org/index_files/farm.htm. Accessed on March 17, 2014.

"The History of Solar." Energy Efficiency and Renewable Energy. http://www1.eere.energy.gov/solar/pdfs/solar_timeline.pdf. Accessed on March 21, 2014.

Holmes, George C. V. 1906. *Ancient and Modern Sailing Ships.* London: His Majesty's Stationary Office. http://www.gutenberg.org/files/33098/33098-h/33098-h.htm#f2. Accessed on March 13, 2014.

Hubbert, M. King. "Nuclear Energy and the Fossil Fuels." http://www.hubbertpeak.com/hubbert/1956/1956.pdf. Accessed on March 20, 2014.

"Illustrated History of Wind Power Development." http://telosnet.com/wind/early.html. Accessed on March 15, 2014.

"The Jacobs Wind Electricity Company." http://www.windcharger.org/Wind_Charger/Jacobs_Wind_Electric_Co.html. Accessed on March 17, 2014.

Kealey, Edward J. 1987. *Harvesting the Air: Windmill Pioneers in Twelfth-Century England.* Berkeley: University of California Press.

"Louis Constantin." IPEXL. http://patent.ipexl.com/assignee/Louis_Constantin_1.html. Accessed on March 20, 2014.

McVeigh, James, et al. 1999. "Winner, Loser, or Innocent Victim? Has Renewable Energy Performed as Expected?" Washington, DC: Resources for the Future. http://www.rff.org/Documents/RFF-DP-99-28.pdf. Accessed on March 19, 2014.

Milborrow, David. "Are Three Blades Really Better than Two?" *WindPower Monthly.* http://www.windpowermonthly.com/article/1083653/three-blades-really-better-two. Accessed on March 19, 2014.

Mooiman, Mike. "Energy in New Hampshire." http://nhenergy.blogspot.com/2013/06/windfall-wind-energy-in-new-hampshire.html. Accessed on March 23, 2014.

Morrison, Hugh. 1952. *Early American Architecture.* New York: Oxford University Press.

"Mr. Brush's Windmill Dynamo." 1890. *Scientific American.* 63(25): 54.

"Mr. Brush's Windmill Dynamo." 1891. *The Electrical Review.* 28: 14.

Mukasa, Alli D., et al. *Development of Wind Energy in Africa.* African Development Bank. http://www.afdb.org/fileadmin/uploads/afdb/Documents/Publications/Working%20

Paper%20170%20-%20Development%20of%20Wind%20 Energy%20in%20Africa.pdf. Accessed on March 25, 2014.

Needham, Joseph, ed. 1965. *Science and Civilization in China,* vol. 4, Physics and Physical Technology. Cambridge, UK: Cambridge University Press.

"New England Wind Forum." Energy Efficiency & Renewable Energy. http://www.windpoweringamerica.gov/newengland/history_windfarms.asp. Accessed on March 23, 2014.

Nixon, Richard M. 1980. *The Real War.* New York: Warner Books.

"October 28, 1980 Debate Transcript." Commission on Presidential Debates. http://www.debates.org/index.php?page=october-28-1980-debate-transcript. Accessed on March 22, 2014.

Parry, Sam. "Reagan's Road to Climate Perdition." Consortiumnews.com. http://consortiumnews.com/2012/01/29/reagans-road-to-climate-perdition/. Accessed on March 22, 2014.

"Petroleum Chronology of Events, 1970–2000." U.S. Energy Information Administration. http://www.eia.gov/pub/oil_gas/petroleum/analysis_publications/chronology/petroleumchronology2000.htm. Accessed on March 20, 2014.

"The Politics of the REA." Farming in the 1930s. http://www.livinghistoryfarm.org/farminginthe30s/money_19.html. Accessed on March 18, 2014.

"Poul la Cour." http://www.poullacour.dk/engelsk/cour.htm. Accessed on March 19, 2014.

Price, Trevor J. "Blyth, James." *Oxford Dictionary of National Biography.* http://www.oxforddnb.com/view/printable/100957. Accessed on March 17, 2014.

Price, Trevor J. 2005. "James Blyth—Britain's First Modern Wind Power Engineer." *Wind Engineering.* 29(3): 191–200.

Project Independence: A Summary. 1974. Federal Energy Administration. Washington, DC: U.S. Government

Printing Office. http://babel.hathitrust.org/cgi/pt?id=uiug.30112028968458;view=1up;seq=15. Accessed on March 21, 2014.

Public Law 93–473. http://www.gpo.gov/fdsys/pkg/STATUTE-88/pdf/STATUTE-88-Pg1431.pdf. Accessed on March 21, 2014.

Public Law 95–91. http://www.gpo.gov/fdsys/pkg/STATUTE-91/pdf/STATUTE-91-Pg565.pdf. Accessed on March 21, 2014.

"R&D Priorities within the Department of Energy." American Physical Society. http://www.aps.org/policy/reports/popa-reports/energy/doe.cfm. Accessed on March 22, 2014.

Ragheb, M. "Wind Generators History." http://mragheb.com/NPRE%20475%20Wind%20Power%20Systems/Wind%20Generators%20History.pdf. Accessed on March 16, 2014.

Ronan, Colin A. 1994. *The Shorter Science & Civilization in China: 4,* an abridgment of the original Needham text (as described earlier). Cambridge, UK; University of Cambridge Press.

Ross, David. 2013. *The Golden Age of Sail.* London: Amber Books.

"Rural Electrification." TVA: Electricity for All. http://newdeal.feri.org/tva/tva10.htm. Accessed on March 18, 2014.

"Savonius Windmill." http://www.creative-science.org.uk/sav.html. Accessed on March 19, 2014.

Shaffer, David. "Xcel to Boost Its Wind Power in Upper Midwest by 33 Percent." *StarTribune Business.* http://www.startribune.com/business/215763441.html. Accessed on March 23, 2014.

Shepherd, Dennis G. 1990. "Historical Development of the Windmill." National Aeronautics and Space Administration. http://wind.nrel.gov/public/library/shepherd.pdf. Accessed on March 15, 2014.

Smith, Joshua M., ed. 2009. *Voyages, the Age of Sail: Documents in American Maritime History. Volume 1, 1492–1865.* Gainesville: University Press of Florida.

Stanford, Dennis J., and Bruce A. Bradley. 2012. *Across Atlantic Ice: The Origin of America's Clovis Culture.* Berkeley: University of California Press.

Stokhuyzen, Frederick. 1965. *The Dutch Windmill.* New York: Universe Books. Also available online at http://www.texva.com/holland/The%20Dutch%20Windmill.htm. Accessed on March 16, 2014.

"Summary of Oregon's Renewable Portfolio Standard." Oregon Department of Energy. http://www.oregon.gov/energy/RENEW/docs/RPS_Long_Summary_July%20 2012.pdf. Accessed on March 23, 2014.

Thomas, R. L. 1982. "DOE/NASA Lewis Large Wind Turbine Program." National Aeronautics and Space Administration. http://ntrs.nasa.gov/archive/nasa/casi.ntrs.nasa.gov/19830006419.pdf. Accessed on March 21, 2014.

"Thomas W. Lawson." *Fleet Sheet.* http://www.fleetsheet.com/lawson.htm. Accessed on March 14, 2014.

"Vertical Wind Turbine v Horizontal Wind Turbine." https://www.youtube.com/watch?v=V2naogzZWF4. Accessed on March 15, 2014.

Vestergaard, Jens, Lotte Brandstrup, and Robert D. Goddard, III. "A Brief History of the Wind Turbine Industries in Denmark and the United States." *Academy of International Business Southeast USA Chapter) Conference Proceedings.* November 2004: 322–327. Available online at http://old-hha.asb.dk/man/cmsdocs/publications/windmill_paper1.pdf. Accessed on March 25, 2014.

Whitewright, Julian. 2009. "The Mediterranean Lateen Sail in Late Antiquity." *The International Journal of Nautical Archaeology.* 38(1): 97–104.

"The Wincharger Corporation." http://www.windcharger.org/Wind_Charger/Wincharger.html. Accessed on March 19, 2014.

"Wind Power from the Darrieus Wind Turbine." EnergyBeta.com. http://www.energybeta.com/windpower/windmill/wind-power-from-the-darrieus-wind-turbine/. Accessed on March 19, 2014.

"Wind Powered Factories: History (And Future) of Industrial Windmills." *Low Tech Magazine*. http://www.lowtechmagazine.com/2009/10/history-of-industrial-windmills.html. Accessed on March 16, 2014.

Wood, A. D., J. F. Ruff, and E. V. Richardson. "Pumps and Water Lifters for Rural Development." http://www.cd3wd.com/cd3wd_40/JF/422/14-363.pdf. Accessed on March 14, 2014.

The World Factbook. Central Intelligence Agency. https://www.cia.gov/library/publications/the-world-factbook/fields/2060.html. Accessed on March 25, 2014.

to ALL
of us!
508-775-9767
NOT FOR
SACRED
.cleanpowernow.org
YES
www.cleanpow
SALE

2 Problems, Controversies, and Solutions

For many decades now, renewable energy has been touted as the answer to our planet's long-term energy needs. To be sure, fossil fuels have served humankind well for over two centuries. However, most analysts believe that, by the very nature of their origins, reserves of coal, oil, and natural gas will not last forever. At some point, humans will have to find new sources of energy, such as solar power, wind energy, tidal movements, wave action, and geothermal activity. These energy resources have the huge advantage that they will probably be available to meet human needs essentially forever. The sun will continue to shine (although not as brightly or as regularly in all parts of the world), the winds will continue to blow (although not as strongly or as dependably), the tides will continue to flow in and out (although with not as much force in all parts of the world), and waves will continue to exist on the open seas (although their action will not always be as accessible to humans). If that line of thinking is true, the question becomes why it is that progress has been so slow in developing these sources of renewable energy.

Wind Power Costs

The capital costs of constructing a wind turbine facility can be broken down into about eight major categories. All but one

---

Supporters and opponents of the Cape Wind project protest outside the U.S. Coast Guard Station in Woods Hole, Massachusetts, on February 2, 2010. Controversy surrounded what would be the first offshore wind farm in the United States. (AP Photo/Julia Cumes)

of these categories relates to the auxiliary facilities needed to build the facility: the foundation on which the turbine is to be built; the installation of electrical connections needed to produce and transport electricity within the wind facility; the grid connection, which includes the power lines leading from the turbine facility to the web through which the electricity will flow; rental of land on which the turbine is to be sited; construction of roads leading to, from, and within the wind facility; the cost of borrowing money needed to design, build, and operate the facility; and consultancy costs. In the average wind facility, these auxiliary costs account for anywhere from 18 to 26 percent of the total cost of the facility. The remaining 74 to 82 percent of costs go to the design and construction of the wind turbine itself ("Cost and Investment Structures" 2009, 98). In other words, the cost of making electricity from wind power depends by far on the way the turbine is designed; it must be built in such a way as to maximize the amount of electrical energy obtained from wind available at the site where the facility is located.

Wind energy has always been expensive, especially during its early years of development in the 1970s, when inventors and engineers had not yet discovered the physical designs that would produce efficient conversion of kinetic (mechanical) energy of moving air into electrical energy. Perhaps more to the point, research developed very slowly at first because most power companies were able to generate electricity from the combustion of fossil fuels at very low cost, because the cost of coal, oil, and natural gas itself was so low. Recall from Chapter 1 that the cost of producing a kilowatt of energy by wind power in 1975 was about 55 cents, compared to a cost of about 5 cents using fossil fuels (McVeigh et al. 1999, Figure 2, page 10; Figure 12, page 22). Under circumstances such as these, private corporations had little motivation to spend the vast sums of money needed on research to produce more efficient wind power systems.

As an example, the first megawatt wind turbine built in the United States was completed, inauspiciously, in August 1941, just four months before the onset of World War II. The machine, commonly known as the Smith–Putnam wind turbine, was built at a cost of $1.25 million in private funds by the S. Morgan Smith Company on a mountainous region known as Grandpa's Knob, near Castleton, Vermont. It was an enormous machine consisting of two blades, each 66 feet in length and 8 feet wide, mounted on a steel tower 120 feet high. It had a rated capacity of 1.25 megawatts (Ragheb 2014; for photos of the Smith–Putnam turbine, see Smith–Putnam Industrial Photos 2014).

The machine operated for about 1,000 hours until a bearing in the gear shaft failed in February 1943. Because of the shortage of materials during wartime, repairs could not be begun until March 1945. Within days after the repairs were completed, one of the turbine's two blades fell off, and the machine stopped working. At that point, the S. Morgan Smith Company decided to discontinue operations, concluding that to do so would only be "throwing good money after bad." The experiment with the wind turbine had been a noble one, but generating electricity by wind power was economically unreasonable at the time (Vermeulen 1974, 58; for a personalized report of this experience, see Putnam 1948).

Government Policies toward Wind Power Research

The Arab oil embargo of 1973 motivated the U.S. government (and the governments of other nations) to begin taking renewable energy sources seriously to an extent never seen before. Within a matter of months, the Congress began to develop strategies for reducing the country's dependence on fossil fuels, in particular the oil imported from Middle East producers. With regard to wind power, those strategies consisted of at least two major approaches. The first approach involved a greatly increased emphasis on the development of wind turbine technology through existing governmental agencies, primarily the National Aeronautics and Space Administration (NASA) and the U.S. Department of Energy (DOE). These agencies provided the funding for new research on turbine technology, although the actual research itself was carried out by private corporations who did not contribute to the financing of the projects. By far the most important component of this research program was the so-called Mod program for research on turbine technology operated by NASA. The emphasis in this program, perhaps reflecting the bias of NASA itself, was on large-scale 3- to 5-megawatt machines. The Mod program eventually accounted for nearly half of all the research funds allocated to wind energy during the 1970s (Norberg-Bohm 2000, 130).

DOE took a somewhat different approach in its support of wind turbine research. It focused instead on smaller wind turbines, with most of the research centered at the Solar Energy Research Institute (SERI) in Golden, Colorado (later renamed the National Renewable Energy Laboratory). Most of this research was also financed by the federal government, although a few of the projects were carried out by private corporations (Norberg-Bohm 2000, 130).

The long-term effect of this research is somewhat debatable. The NASA projects, focused on large wind machines, are generally thought to have been less than successful; no machines of that design were built in the succeeding two decades, largely because NASA had misjudged the willingness of private businesses to build such turbines. The SERI research was somewhat more productive as a number of specific wind technologies developed through these projects eventually found their way into later wind turbine design (Norberg-Bohm 2000, 131).

The U.S. Congress also adopted a number of bills during the 1970s designed to promote the development of renewable energy. Perhaps the most significant of these bills was the National Energy Act of 1978 (NEA). This act consisted of five major parts dealing with industrial fossil fuel use, natural gas, energy conservation, tax policies, and public utilities policies, now constituting Public Laws 95–617 through 95–619. Of these five statutes, the two most related to renewable energy were the Public Utility Regulatory Policies Act (PURPA; P.L. 95–617) and the Energy Tax Act (ETA; P.L. 95–618) (Jones 1980, 325, footnote 7; 326–332; 334–335).

Public Utility Regulatory Policies Act of 1978 (PURPA)

PURPA was important because it struck at the very heart of the electrical power industry in the United States. Prior to 1978, electrical utilities held a monopoly on electrical energy. They controlled every aspect of the production, distribution, and consumption of electricity at every level. PURPA changed that situation by requiring utilities to purchase electricity from small private companies, no matter how that electricity was generated, at the full *avoided cost,* the amount it would have cost a utility itself to produce the electricity. For the first time, anyone who generated electricity by whatever means (including solar, wind, geothermal, or other alternative means) could sell his or her electricity at full market price. Producers of electricity with renewable energy at last had the opportunity to make a reasonable profit on their enterprise, a significant stimulus for the development of new renewable energy facilities ("Public Utility Regulatory Policy Act" 2014; for the text of the act, see "Public Utility Regulatory Policies Act of 1978" 2014).

Many observers argue that PURPA served an essential function when it was first passed in 1978. Small companies generating electricity with the use of renewable fuels had little or no market for their product. Opening up the electric grid to such companies provided an important impetus to the renewable energy field. Over time, that situation has changed. Renewable energy companies have become more successful, and the nature of mainstream electricity-generating companies has changed considerably. Today, PURPA remains in force, although its policies and programs are carried out to a large extent by the states rather than by the federal government. Its most important impact is now in the field of electricity cogeneration rather than the use of specific renewable fuels ("Public Utility Regulatory Policy Act" 2014; Richardson and Kaufmann 2014).

Energy Tax Act of 1978

The Energy Tax Act of 1978 (ETA) had two primary objectives: to reduce the consumption of traditional fuels in the United States, while encouraging the development and use of alternative, renewable, or otherwise "unconventional" forms of energy, such as solar, wind, geothermal, biomass, and tidal. The act attempted to achieve these objectives using tax incentives and disincentives. Specifically, it reduced or eliminated some of the tax advantages previously provided to energy companies for their use of coal, oil, and natural gas in their operations. It also imposed a so-called gas-guzzler tax on the sale of vehicles with low gasoline use efficiency.

On the other hand, the act provided a number of tax incentives for the conversion of energy operations from fossil fuel operations to the use of renewable resources. A tax incentive is some feature of tax policy that encourages individuals and businesses from carrying out some types of activities and/or discourages them from carrying out other types of activities. It also offered tax breaks for new research on and development of alternative energy sources. For example, the act provided for a somewhat complicated calculation of a tax credit for the use of renewable energy equal to the sum of (1) 15 percent of the energy conservation expenditures up to a maximum of $2,000 and (2) 30 percent of qualified renewable energy source expenditures for solar, wind, and geothermal energy equipment but not to exceed $2,000 plus 20 percent of such expenditures up to $2,000 but not exceeding $10,000 (Bill Summary & Status 2014; for a more detailed analysis of the act's provision for wind energy, see Hinman 2006; see also Righter 1996, 99).

Federal Renewable Energy Policies since 1978

The history of federal policies with regard to the development of renewable energy sources since the late 1970s is complex, irregular, and often confusing. The increased emphasis on alternative and renewable fuels reflected in the adoption of the NEA. With the election of President Ronald Reagan in 1980, a new (actually a renewed) energy policy emphasized the role of fossil fuels, rather than renewable fuels, in the nation's future. (For a review of Reagan's energy policy, see "The Executive Branch and National Energy Policy: Time for Renewal" 2012.) Many of the energy programs outlined in the NEA scheduled to expire in the 1980s were allowed to do so, business tax credits in 1982, and residential tax credits in 1985. Only select provisions of the 1978 law dealing with renewable property were extended (Sherlock 2014, 4).

A shift in tax policy occurred once more with the adoption of the Energy Policy Act of 1992 (EPA). In the aftermath of the Reagan and George H. W. Bush administrations' emphasis on the development of fossil fuel resources, the U.S. Congress once more began to envision the use of tax credits as a mechanism for encouraging the development of renewable energy resources. A major difference in the 1992 act, in comparison with the 1978 act, however, was the decision to base tax credits on the actual *production* of renewable energy in contrast to the *installation* of energy facilities. Individuals and companies were now to receive tax credits when their facilities actually began to generate electricity from wind, solar, geothermal, or other renewable sources rather than for just building those facilities, whether they efficiently produced electricity from them or not (Logan and Kaplan 2014, footnote 3, page 2).

The U.S. Environmental Protection Agency (EPA) was a comprehensive piece of legislation, covering a wide variety of topics, including energy efficiency, natural gas, electric vehicles, radioactive waste, uranium enrichment, coal, the Strategic

Petroleum Reserve, global climate change, oil pipelines, environmental effects of energy use, hydroelectric power, nuclear power production, and alternative and renewable fuels ("Energy Policy Act of 1992" 2014). The act provided for a tax credit based clearly on the actual production of electricity by a renewable energy facility ("based on the number of kilowatt-hours of electricity generated by the facility through the use of solar, wind, biomass, or geothermal energy"; "Energy Policy Act of 1992" 2014, 216) The amount of the tax credit to be given for such a facility was 1.5 cents per kilowatt-hour, although the secretary of energy was authorized to adjust such payments in such a way as to maximize the incentive based on the financial resources available to the DOE. Interestingly enough, the tax credit provided for in the EPA applied originally only to wind energy and to energy obtained from biomass, although the provision was later broadened to include other forms of renewable energy, such as solar, geothermal, and hydroelectric power ("Energy Policy Act of 1992" 2014, 272; Sherlock 2011, 4–5).

The original production tax credit (PTC) authorized by the EPA expired in July 1999 but was reauthorized in December of the same year by the Ticket to Work and Work Incentives Improvement Act of 1999 and then again in March 2002 by the Job Creation and Worker Assistance Act of 2002, once more by the American Jobs Creation Act of 2004, again by the Energy Policy Act of 2005, and again by the American Recovery and Reinvestment Act of 2009. Most recently, tax credit provisions for wind and other renewable forms of energy were reauthorized by the American Taxpayer Relief Act of 2012. The tax credit for wind-generated electricity is now 2.3 cents per kilowatt-hour ("Renewable Electricity Production Tax Credit [PTC]" 2014).

The EPA enacted a second type of tax incentive for wind power in 1992 with the creation of the Renewable Energy Production Incentive (REPI). This program involved a cash payment to producers of electricity who used some form of renewable source for their activities. The original payment under the REPI program was 1.5 cents per kilowatt-hour. That program was reauthorized in 2005 and scheduled to expire in 2026. The program lapsed, however, when it was no longer funded by the federal government, and many of its activities were taken over by comparable state laws in the second decade of the twenty-first century ("Renewable Energy Production Initiative" 2014; for an example of a state program, see "Maine: Incentives/Policies for Renewables & Efficiency" 2014).

State Programs

While the federal government has slowly and haltingly moved forward over the past half century in its efforts to develop on ongoing policy about tax credits and other financial incentives for the promotion of wind energy, most individual states and some localities have also moved in the same direction, often with greater effectiveness and greater consistency. Currently, the incentives offered by states and municipalities fall into about ten major categories: personal, corporate, sales, and property tax credits; rebates; grants and loans; industry support; bonds; and performance-based incentives. (For a review of these categories, see "Financial Incentives for Renewable Energy" 2014.)

The range and variety of programs offered by the states are considerable. They range from as few as 3 different programs in Kansas to as many as 74 different programs in California, which has long been the leader in state wind programs in the nation. As an example of the programs available, consider some of the programs available in only one state, Massachusetts. One such program is a property tax incentive that exempts a utility from all property taxes for a period of twenty years for land used for the generation of solar, wind, hydroelectric, and certain other types of renewable energy. A second state program is the Commonwealth Wind Incentive Program—Micro Wind Initiative, which pays a rebate of $5.20 for public projects ($4 for nonpublic projects) per watt for systems that produce less than 100 kW of energy, with a maximum rebate of $130,000 ($100,000 for nonpublic projects). Massachusetts also allows both a corporate and a personal excise tax deduction for income derived from the sale or lease of a patent on any renewable energy device or system (all information obtained from "Massachusetts: Incentives/Policies for Renewables & Efficiency" 2014).

Effects of Tax Credits on Wind Power

The variability of programs in governmental support for renewable and alternative energy since the 1970s has been, to a considerable extent, a reflection of differing political philosophies about the appropriate roles of the public and private sectors in the support of basic research. (Basic research is defined as research to answer fundamental questions about some aspect of the natural world without concern for any potential practical applications of that research.) Some people argue that large-scale basic research projects can be carried on *only* by governmental agencies, usually the federal government. They point to examples such as the Manhattan Project for the development of the first nuclear weapons, the Apollo space program designed to place the first human on the Moon, the ill-fated Superconducting Super Collider designed to study the basic structure of the atom, and the Human Genome Project (HGP), whose purpose it was to unravel the complete genomic structure of humans. Such projects, they say, are far too risky and much too expensive for any portion of the private sector to take on (Kennedy 2012; Singer 2014).

Other people suggest that that position does not give proper credit to the power of the marketplace in which individuals and companies compete to achieve even the loftiest of research goals. They may point to the example of the Bell Laboratories, which, until 2008, produced some of the most important scientific advances in modern history, such as the discovery of the wave nature of matter, the invention of the transistor, and the electronic structure of magnetic materials (Ganapat 2008). The proper role of government, they argue, is to "get out of the way" of private enterprise and let business attack even the most difficult of research challenges. (In recent years, a number of legislators, industrialists, commentators, and other observers have suggested this view and used exactly this phrase. See, for example, Johnsen 2014; "Rep. Paul Broun" 2014; Summary of Hearings on Energy 2014).

As might be expected, this debate is hardly clear-cut. For example, while the federal government led the long-term and expensive effort to unravel the human genome in the HGP, that project eventually attracted the attention of an ambitious private company, Celera Genomics, which eventually contributed almost as much to the success of the HGP as did the federal effort itself (Philipkoski 2014). Also, while the federal government was largely responsible for the ultimate success of the Apollo program, it has essentially ceded the lead role for similar research in the early twenty-first century to a small group of private companies (Enke 2014).

In any case, the difference of opinion between the appropriate role of private and public sectors in the development of renewable energy sources has played out across the past half century, with first one view and then the other prevailing. During the Carter administration, for example, research on solar, wind, geothermal, and other unconventional energy sources was largely the purview of federal and state governments. Under President Ronald Reagan, however, that situation was reversed, and the federal government essentially handed over virtually all research on renewable technologies to the private sector.

In the end, the most important question might well be, which approach has been more successful in the development of alternative energy sources: the support of federal and state governments or the work of private corporations? The question is important because it may suggest the most appropriate pathway to take in the future in the promotion of unconventional energy sources. A good deal of hortatory writings has been devoted to this topic, and some research has been conducted on the question. Some research seems to suggest that the loss of tax incentives for the production of wind energy has had a significant, and often dramatic, effect on the development of that field since the 1990s. In a survey by the American Wind Energy Association conducted in 2013, for example, researchers found that the expiration of the PTC on four occasions resulted in a drop in wind capacity installation in the following year. Expiration of the PTC in 1999 resulted in a decrease in wind capacity installation by 92 percent in 2000; expiration of the tax in 2001 resulted in a decrease by 76 percent in 2002; expiration of the tax in 2003 resulted in a decrease by 76 percent in 2004; and expiration of the tax most recently in 2012 resulted in a decrease in wind energy installation by 99 percent in 2013 ("Federal Production Tax Credit for Wind Energy" 2014). A somewhat earlier study by the U.S. Energy Information Administration found essentially the same results. An extended discussion of the effect observed in this study can be found at http://www.eia.gov/todayinenergy/detail.cfm?id=8870 (see also "Production Tax Credit for Renewable Energy" 2014).

Some research, however, has produced very different findings from those summarized earlier. For example, a study of the effect of various state tax credit policies for the development of wind energy facilities in those states found that there was essentially no evidence that state policies had any beneficial effects on the development of wind energy. The author of the research concluded that "I find almost no evidence to suggest that these credits increase wind development" and that, further, "I conclude that the efficacy of pro-wind policies depends largely on the ratio of wind resources to solar, hydro, and other renewable resources in a given state" (Indvik 2010, 1).

Perhaps more important than the research that has been done on the relationship of tax credits to wind power development is the political debate over this topic. Proponents of wind power argue that the technology is only one, but an important one, of the elements needed for a robust renewable energy future for the United States. The country (and the world) cannot continue to rely on fossil fuels forever, and efforts must be begun and sustained to find ways of making use of the wind power freely available on the planet. The problem is that, like many types of research, private industry has not been willing to invest in basic research and the development of wind energy; only the federal and state governments have been willing to do so, through tax credits and other tax incentives. If the development of wind power in the United States is to continue, it will do so only with the support of some forms of tax incentives, such as tax credits. (For a general overview of the arguments in support of wind energy development, see "Why Wind Energy?" 2014; for arguments in support of wind tax credits, see Schweiger and Kiernan 2014).

Opponents paint a different picture of tax credits for wind energy. They tend to view tax credits for renewable energy as a "handout" by the government to a technology that cannot make it on its own. They argue that all forms of energy development should be able to survive on their own in the marketplace without special subsidies to keep them alive. One of the most outspoken and influential of the anti-tax-credit groups for wind energy is an organization called the American Energy Alliance (AEA). AEA's mission statement says that the organization's main purpose is "to enlist and empower energy consumers to encourage policymakers to support free market policies." It goes on more specifically to suggest that

> free markets will provide the United States with affordable, plentiful, and reliable energy. Energy consumers, not bureaucrats, should decide the mix between various sources of energy. The tax code should not be used to pick energy winners and losers. Lastly, markets not mandates result in lower energy prices and more abundant energy for all Americans. ("Mission Statement" 2014)

In September 2013, the AEA joined with twenty-two other conservative and free-market organizations to ask Congress not to renew the wind energy tax credit scheduled to expire at the end of the year. The letter claimed that the tax credit "doesn't produce cheaper energy, it threatens electrical grid reliability, it's inefficient, it's unprincipled tax policy, . . . and it's time to end this misguided handout" ("Coalition to Congress" 2014). Among other signers of the letter were the American Coalition for Clean Coal Electricity, the American Enterprise Institute, Americans for Prosperity, the Cato Institute, the Center for the Study of Carbon Dioxide and Global Change, the Coalition for Affordable American Energy, the Competitive Enterprise Institute, the Congress of Racial Equality, Freedom Works, the Heartland Institute, the Heritage Foundation, Koch Industries, the National Black Chamber of Commerce, the Tax Foundation, and the U.S. Chamber of Commerce. A major contributor to the anti-wind campaign is reported to have been Koch Industries, active in a number of current energy and other political issues (for more information on these groups, see "Who's Behind the Smears?" 2014).

Another expression of the opposition to PTCs for renewable energy came in a budget proposed by U.S. representative Dave Camp (R-MI), chair of the House Ways and Means Committee. In his comprehensive 2014 bill designed to reform the U.S. tax code, Camp recommended the repeal of essentially all tax credits and incentives designed to encourage the development of all forms of renewable energy ("Discussion Draft 2014," Title I, Subtitle D; Title III, Subtitle C). The president of the AEA, Tom Pyle, commended the draft document for recommending the cancellation of all tax credits for renewable energy. He said that Camp had taken an "encouraging step" forward in eliminating "wasteful green energy subsidies, including the wind PTC" ("Camp's Tax Reform Plan" 2014). As noted earlier, the Congress decided instead to continue the tax credit for wind energy in 2014.

Proponents of tax credits for renewable energy often counter the arguments presented here by pointing out that federal tax credits favor the production and use of fossil fuels far more than they do renewable resources. A 2009 study by the Environmental Law Institute found that federal subsidies to fossil fuels in the period 2002 to 2008 amounted to $72.5 billion, whereas subsidies to renewable fuels (not including ethanol) amounted to $12.2 billion, a sixfold margin of fossil over renewable fuels. In addition, the study found that funding of fossil fuels increased significantly during the period of the study, whereas funding of renewables experiences a significant decline (*Estimating U.S. Government Subsidies to Energy Sources: 2002–2008* 2014; a graphical representation of these data can be found at http://www.eli.org/sites/default/files/docs/energy_subsidies_black_not_green.pdf.).

Wind Farms and Land Use Issues

The use of tax incentives to promote the development of wind energy is by no means the only issue that surrounds discussions of that type of renewable energy. Another issue that often stirs debate centers on the land area needed to develop large-scale wind projects. One way to envision the space needed for a large wind farm can be gained from a review of what is currently the largest wind farm in the world, located in the Altamont Pass region of northern California, about 50 miles east of San Francisco. The region currently has more than 4,000 wind turbines, most of which were installed in the 1970s when federal and state tax policies were most favorable to the development of wind farms. (Statistical data on the Altamont and other wind farms often differ depending on the source from which they come.) These wind turbines cover an area of about 78 square miles and produce about 511 megawatts of electricity (Chapter 2 2014, 2–1; "Fast Facts about California Wind Energy" 2014).

This view of the space needed for a commercial wind farm is somewhat misleading, however, as the actual space required for any given facility depends on a number of different factors. Strictly from the standpoint of the wind turbine itself, a general rule of thumb is that each turbine must be surrounded by a distance equal to anywhere from 5 to 15 rotor diameters. The rotor diameter of wind turbines today ranges from about 15 meters (50 feet) to more than 100 meters (300 feet), so large wind turbines on a wind farm might have to be as much as 4,500 feet from each other ("Environmental Impact of Wind Farms 2014"; "Optimizing Large Wind Farms" 2014). In addition to the space required for the turbines themselves, space is also needed for ancillary materials and structures, such as power lines and access roads. These requirements mean that relatively few areas in metropolitan regions are available for the construction of wind farms, and those that are built essentially commandeer all of the space available. That fact explains how it happens that nine of the ten largest wind farms in the United States are currently located in remote regions of two of the nation's states with the most open spaces—Texas and California (Madrigal 2014). In sum, predicting the size of a wind farm is a difficult challenge and can vary considerably depending on local factors ("Wolds Wind Farm Opposition" 2014); most such farms do take up a very large land area).

Offshore Wind Farms

It is largely for this reason that a movement began in the early 1990s to begin building wind farms not on land but on the open water. The first such wind farm was built off the north coast of the Danish island of Lolland adjacent to the village of Vindeby in 1991. It consisted of just eleven wind turbines with a rated capacity of 4.95 megawatts. The farm was originally designed as a pilot project to determine the efficacy of offshore wind farms, but it has continued to operate well ever since. Research studies have found that the farm produces about 20 percent more electricity than would comparable onshore farms of the same size and general location. They have also found no detrimental effects to the environment ("Offshore Center Danmark" 2014, 34). The success of the Vindeby project has had a significant effect of the Danish government's later decisions to expand the use of offshore wind power for the production of the nation's electrical needs. (An interesting video showing the construction and operation of offshore wind turbines is available at https://www.youtube.com/watch?v=839W90ZOEWY.)

The success of early offshore wind farms like that near Vindeby changed the metrics of wind farm construction, leading to the building of more than 1,000 such facilities in the following fifteen years. Today, a single wind turbine on an offshore farm usually produces more power than all eleven of the original turbines at Vindeby. The largest offshore wind farms in operation or under construction as of late 2014 were the London Array, located at Kentish Knock, in the North Sea, off the east coast of the United Kingdom (175 turbines with a rated capacity of 630 MW), Greater Gabbard, also in the North Sea (140 turbines; 500 MW); Trianel Borkum and BARD Offshore 1, both in the North Sea, north of the German coast (each with 80 turbines and each rated at 400 MW); and the Anolt Offshore Wind Farm, in the ocean between Denmark and Sweden (111 turbines; 400 MW).

The United Kingdom is currently the world's leader by far in the development of offshore wind farms. At the end of 2012, the year for which recent data are available, the nation had a rated capacity of 2,947.9 megawatts of offshore wind facilities, just over half (54.3 percent) of the world's total capacity. The nation with the second-largest offshore capacity was Denmark (921.0 MW; 17 percent of world capacity), followed by China (389.6 MW; 7 percent), Belgium (379.5 MW; 7 percent), and Germany (280.3 MW; 5 percent). The only other nations with offshore wind farms were the Netherlands, Sweden, Finland, Japan, Ireland, Spain, Norway, and Portugal ("2012 Annual Report," 9).

As of late 2014, there were no offshore wind farms in the United States, although a number of such facilities had been in the planning stages, some for more than a decade ("7 Future Wind Farms" 2014). The project closest to realization as of late 2014 was Cape Wind, which actually began construction in December 2013, thus qualifying for federal tax credits for wind energy that were due to expire at the end of the year. "Getting in under the wire" on the project was important for Cape Wind because it meant the company would qualify for as much as $780 million in cash reimbursement from the federal government for beginning construction on a wind project (Goossens 2014).

It had taken Cape Wind a very long time to reach this point. The project was first proposed in 2001 by Cape Wind Associates. It was originally to consist of 170 turbines (later reduced to 130 turbines) with a rated capacity of 3.6 megawatts each in an approximately 46-square-mile portion of Nantucket Sound, within sight of Nantucket Island ("Cape Wind" 2014). The project had a long and complex history partly because of objections raised by a number of groups and individuals expecting to be affected by the project and partly because government oversight of the project was transferred from agency to agency. An initial court case by objectors was settled in 2002, when the court ruled that Cape Wind could go ahead with initial plans and testing. The company then began the long process of testing equipment, conducting environmental impact studies, and attempting to placate opponents.

The first important step forward occurred in 2004 when the U.S. Army Corps of Engineers, then charged with oversight of the project, issued a favorable environmental impact statement (EIS), essential for continuation of the project. That decision was countermanded a year later, however, when the EPA ruled that that EIS was inadequate and that further environmental studies were needed. In the ensuing years, individual steps forward on the project were approved and completed, such as the laying of transmission lines in 2005 and approval of the project by the state department of energy and the environment in 2007. Finally, in 2010, Cape Wind was issued a formal lease for the construction and operation of a wind facility in Nantucket Sound (for a history of the project, see "Cape Wind" 2014; Daley 2014).

Even this action did not resolve the issue, however, as objections continued to arise from local communities affected by the project, the Federal Aviation Administration, and some Native American tribes in the area. Finally, on April 28, 2010, Interior secretary Ken Salazar announced his approval of the Cape Wind project, all of which did not mean that the project could begin work the next day since it still had not arranged financing for the project. Meeting that hurdle took another three years, and the project was at last able to begin construction at the end of 2013.

Objections to Wind Farms

The objections to the construction of a wind farm in Nantucket Sound by Cape Wind provide a glimpse of the concerns that many individuals have about the development of wind projects in general, whether they take place on land or offshore. Of course, placing wind turbines over the water mitigates one of the major objections to the development of wind power, namely, the harm, dangers, and/or disadvantage of having dozens or hundreds of wind turbines spread out over a few dozen or a few hundred square miles of valuable land. But a number of other objections remain, whether turbines are placed on land or over the water. (One of the most comprehensive reviews of the range of objections raised to wind power generation, with responses to those arguments, is Clarke 2014.)

One of the most common concerns expressed by residents of the Nantucket Sound area has been that the Cape Wind project would destroy a pristine and beautiful seascape that has brought great pleasure to untold numbers of visitors and permanent residents. An expression of this view is provided by the

Web page of Save Our Sound (SOS), an organization created to oppose the Cape Wind project. The page claims that the wind turbine project will "dramatically alter the natural landscape and negatively impact several historic landmarks." At night, SOS goes on, the sea will look like New York City's La Guardia airport with flashing red and amber lights and foghorns ("The View" 2014; the Web page also has artists' conceptions of the view of the area after installation of turbines).

One of the best known and most ardent critics of Cape Wind was Senator Edward Kennedy, whose family home in Hyannisport overlooks the proposed site of the wind farm. Kennedy and his family complained that the farm would irreparably damage their view of Nantucket Sound and would pose a hazard to boaters, like themselves, who used Nantucket Sound for recreational purposes. After Senator Kennedy's death in 2009, other members of his family continued to raise objections to the Cape Wind project. In 2011, for example, Senator Kennedy's nephew, Robert F. Kennedy, Jr., asked Secretary Salazar not to approve the project, although the basis for his objections differed from that of his uncle. He said that the project was economically unsound, calling on taxpayer funding for its success. It was, he was quoted as saying, "a boondoggle of the worst kind" (Seelye 2014). As an environmental lawyer and a proponent of the development of renewable energy, Kennedy's comments aroused special attention. Some observers saw hypocrisy in his taking a "not in my backyard" (NIMBY) view of a technology that he might otherwise have supported in a different setting (see, for example, Ransom 2014).

Aesthetic Concerns

The dispute over the construction of wind turbines in Nantucket Sound highlights the fact that aesthetic concerns about wind power are often the most commonly heard complaints about wind power projects (National Research Council 2007, 10). The somewhat striking fact is that, when people are asked about their feelings about wind power *in general,* they tend to express strong support for the technology, reflecting their beliefs that greater emphasis needs to be placed on the development of renewable energy sources. Such surveys commonly find support for wind power as high as 75 percent of the general population ("As Gas Prices Pinch" 2014; Jacobe 2014). Yet, when they are asked about specific wind energy projects planned for their immediate area, their support tends to drop, often quite dramatically. Studies seem to suggest that individuals view a wind farm *in their own neighborhood* significantly more negatively than they do wind energy in general. In some studies, support of wind power drops by 20 percentage points or more when respondents learn that a wind farm is planned for their own neighborhood (Smith and Klick 2014).

Researchers often explain this difference by invoking the principle of "not in my backyard". According to this principle, individuals often view a technology (e.g., wind power) quite differently when they realize that it will be installed adjacent or close to the place where they live or work. At that point, some of the disadvantages of that technology become more than just abstract issues; they become everyday realities. Aesthetic issues are especially potent NIMBY factors because all one has to do in regard to a wind farm is to look out of the front door and see dozens of wind turbines, often day and night, blocking or disrupting a particularly beautiful view that may even have been a factor in a person's buying or building a home.

Certain businesses may also be (or think they are) adversely affected by the installation of a wind farm. For example, tourism companies and agencies may be under the impression that a new field of wind turbines will spoil the natural beauty of an area, discouraging tourists from visiting the area. The aesthetic effects of a wind farm may be more than a psychological concern then; they may also have an economic effect on an area. A March 2013 article in the *Berwickshire News* (England), for example, pointed out that five recent studies about the effect of wind farms on tourism had all reached the same conclusion: such facilities caused a reduction in tourism in regions where the farms were located ("Tourism Blown Off Course by Turbines" 2014).

That view may be somewhat skewed, however, in that it assumes that landowners and tourists are *all* offended by the view of a wind farm. In fact, there is good evidence that at least some individuals actually view such facilities favorably. As one graphic designer has written:

> I must confess . . . to finding wind farms in general beautiful. . . . For many of the same reasons I find minimalist sculpture, the International Style in architecture, and modernist graphic and industrial design in general appealing visually, I find the wind farm visually appealing. (Good 2006, 82)

Research also suggests that opposition to wind farms tends to diminish to some extent after such facilities are built and neighbors discover that the disadvantages about which they were most concerned may not actually be as offensive as they had imagined when the project was first announced (Damborg 2014).

Given that individuals almost inevitably differ in their views as to what is aesthetically pleasing in art, architecture, natural landscapes, and other environmental situations, it is probably to be expected that they will also differ with regard to their views on wind farms. But a number of observers have pointed out that planners can probably do a much better job of reducing the conflict between pro- and anti-farms in the way they approach new projects. One review of this issue points out that some countries consistently realize a lower degree of conflict in the siting of wind farms than do others for recognizable reasons. These countries (the review cites Denmark and Germany as examples) make a special effort to involve local communities and individuals, who are likely to be affected by new wind sites, in planning and decision making about wind projects. In contrast, countries that are less likely to do so (the United Kingdom and the Netherlands are mentioned as examples) tend to experience greater conflict over the construction and operation of wind farms (Wolsink 2007; see also National Research Council 2007, Chapter 5).

Noise Issues

Closely related to the issue of visual aesthetics associated with wind farms is the potential problem of excessive noise associated with the operation of turbines. A search of the Internet will reveal a very large number of complaints from individuals who live near wind farms and claim that wind turbines produce an excessive level of noise. Measuring the precise noise level produced by any one turbine or any one farm can be difficult because so many different factors are involved. Noise from the turbine itself can usually be attributed to two major sources: the mechanical sound produced by devices within the nacelle, such as the moving gears and rods, and the "swishing" sound of the blades themselves. But the sound produced by a single wind turbine in isolation from its surroundings is only a portion of the sound a nearby resident might hear. Other factors that are involved are other environmental sounds, such as the sound of the wind itself, the roar of a river, or the lapping of waves, as well as sound emitted by human sources, such as road and rail traffic, the hum of nearby machinery, and the sounds of human voices.

A number of attempts have been made to quantify the amount of noise produced by various types of wind turbines, although the precise conditions under which such measurements were made have not always been provided. For example, the General Electric company (GE) has published a graph of the amount of noise that can be heard at various distances from one of its wind turbines, presumably one of the 1.5 MW turbines it manufactures. The noise level at the base of the turbine itself, according to this graph, should be about 105 dB (decibels), comparable to the noise produced by a residential lawnmower. At a distance of 100 meters (300 feet), the noise level is about 50 dB, comparable to the sound produced by an in-window air-conditioning unit. At a distance of 300 meters (about 1,000 feet), the closest approach a wind turbine is normally placed to a personal residence, business, or some other structure, the sound level is about 45 dB, between that of the air conditioner and an average household refrigerator. (This graph is available online at http://files.gereports.com/wp-content/uploads/2010/11/larg-wind-turbine.jpg.)

Data about noise levels from wind turbines can vary significantly from those shown in the GE publication. A review of the literature on the topic by Jim Cummings of the Acoustic Ecology Institute notes that "actual sound contour maps generated by consultants hired by wind energy developers paint a much less sanguine picture" than that offered by the GE graph. A summary of four actual wind farms for which noise data have been collected shows that the distance at which turbine noise drops to 45 dB ranges from about 360 meters (1,200 feet) to as much as 900 meters (3,000 feet) (Cummings 2012, 13).

Possibly as important as any raw scientific measurements of sound levels produced by wind turbines is the response that such noise produces in nearby residents. Opponents of wind power have assembled a number of anecdotal comments by such individuals describing their own personal experiences with turbine noise, such as:

- As a neighbor of Hoosac Wind living 1 mile from 10 GE 1.5 MW turbines I can say conditions at the project can be loud enough to wake me from sleep.
- Like Chinese water torture; the thumping sound of turbines will drive a lot of people "crazy," esp. at night.
- The vibrations (aka "noise") from a wind turbine may be much less harmful than radiation from a leaking nuclear reactor, but it is still a nuisance.

- The noise is pervasive, it impacts ones health, destroys a good night's sleep, devalues ones home, and generally makes everyday life a challenge for those who live within a half a mile (or more) of industrial sized wind turbines.
- OK, it may not give you cancer, heart disease, or diabetes, but if it's very disturbing to you and your family, then it may have psychological impact and THAT is a health issue.
- [It] is the unheard infrasound component of wind turbine noise that causes problems to nearby residents. (All quotes from Casey 2014; "Measuring Wind Turbine Noise" 2014)

Infrasound

The last of these comments refers to a specific type of effect produced by the operation of a wind turbine in turbulent air. The term *infrasound* refers to very low frequency sounds that are beyond the limits of human perception. That is, a wind turbine may produce, in addition to the noises that trouble many individuals, another form of sound that is normally not detected by the human ear and that may be related to health problems. One of the most outspoken proponents of this line of research is Alec N. Salt, a professor in the Department of Otolaryngology and Head and Neck Surgery at Washington University School of Medicine, in St. Louis. Salt has argued that the fact that the human mind does not perceive infrasound does not mean that it poses no threat to human health. Instead, he suggests that infrasound may result in a variety of biological changes that express themselves as a variety of symptoms, including annoyance, stress, sleep disturbance, panic, chronic sleep deprivation, blood pressure elevation, memory dysfunction, unsteadiness, disequilibrium, vertigo, nausea, "seasickness," tinnitus, and sensation of pressure or fullness in the ear (Salt 2014).

Another active proponent for the case that wind turbines may have identifiable effects on human health is physician

Nina Pierpont. Pierpont has had experience both in the academic field (she was clinical assistant professor of pediatrics at Columbia University's College of Physicians and Surgeons) and in the medical field (she was a practicing physician, working at a native-run hospital on the Alaskan tundra). Pierpont explains that she was first made aware of the possible health hazards of wind turbines, which she has termed wind turbine syndrome (WTS), while she was working as a country doctor and began receiving complaints from patients of a number of disorders that she eventually attributed to infrasound produced by wind turbines in the area. She went on to write a book on the topic (Pierpont 2009) that has become the focus of a vigorous back-and-forth debate over the possible health effects of wind turbines on human health(see, for example, Layton 2014).

Individuals who are convinced of the reality of a WTS effect point to a large collection of anecdotal reports as well as some research studies that they say support their views. For many such individuals, the "smoking gun" in the case against wind turbines is a research study produced by the U.S. DOE in 1987. That study, they say, proved that wind turbine noise is "real, not imaginary" and that it has the potential for causing human health problems (Delingpole 2014; the DOE study can be found at http://docs.wind-watch.org/Kelley_Proposed-metric-assessing-potential-annoyance-wind-turbine-LF.pdf).

As might be expected, other individuals are equally convinced that infrasound poses no real risk to human health. For example, a longtime critic of the WTS hypothesis, Dr. Simon Chapman, professor of public health at Sydney University in Australia, has studied the purported effects of wind turbine sounds on nearby residents of wind farm facilities. In 2013 he published a report on some of this research, which, he says, shows that "the evidence for wind turbine noise and infrasound causing health problems is poor." Instead, he suggests, complaints about supposed health effects from wind turbines are more likely to be psychological effects resulting from the exchange of anecdotal reports among individuals who live close to wind farms (Chapman et al. 2013).

The issue of health effects from wind turbines remains a very active field of research and disagreement as of late 2014. Some industry groups and medical associations have expressed serious doubts as to whether turbines produce infrasound at a level that can be harmful to the human ear. In March 2014, for example, the Australian Medical Association issued a position statement concluding that "[t]he available Australian and international evidence does not support the view that the infrasound or low frequency sound generated by wind farms, as they are currently regulated in Australia, causes adverse health effects on populations residing in their vicinity" ("Wind Farms and Health—2014"). In contrast, the Society of Rural Physicians of Canada published its own review of the research on infrasound also in 2014 and concluded that "[i]naudible low-frequency noise and infrasound from IWTs cannot be ruled out as plausible causes of health effects" (Jeffrey, Krogh, and Horner 2014, 21; for an extended review of the research on infrasound and other health effects of turbine noise, see Colby et al. 2009; Farboud, Crunkhorn, and Trinidad 2013; Knopper and Ollson 2011).

More about Noise

As the comments discussed earlier suggest, noise is often cited as a problem for people living near wind farms. But a number of studies have shown that the problem is a good deal more complicated than it may seem at first glance. For example, consider an experiment in which a group of people are placed in a situation where they have an opportunity to hear noise from a variety of sources: automobile traffic, construction projects, urban street sounds, and wind turbines, for example. In such an experiment, one might predict that the subjects in the experiment would react in much the same way to the same level of noise from all sources. That presumption turns out not to be true. In an experiment conducted with 725 individuals in the Netherlands, researchers found that subjects became more annoyed by the sound of wind turbines, even when the noise level was the same as that from traffic and industrial operations.

The researchers hypothesized and tested a variety of factors that might explain this result. They discovered that, for example, simply being able to see the wind turbines apparently made the sound from them more annoying. They also determined that the unique sound of the wind turbines (the "swooshing" sound they make) increased the discomfort level of listeners. And, perhaps most interesting of all, they found that people who had negative attitudes toward wind power to begin with turned out to be more annoyed by turbine sounds than those who did not have such feelings. Finally, the researchers discovered that people who, for some reason or another, benefited economically from the operation of wind turbines were less annoyed by the sounds they made. In a final observation, the researchers noted that these results were very similar to those obtained from a similar study conducted in Sweden on a previous occasion (Janssen et al. 2011; Pedersen et al. 2009).

Concern about the noise associated with wind turbines is obviously a problem about which wind companies must be concerned. Those companies, like most businesses, want to be on good terms with their customers and their neighbors. One of the questions wind companies might need to ask about their projects is, "How much noise is too much?" That is, are neighbors of a wind farm willing to put up with 30 dB of noise? Or 40 dB of noise? Or 50 dB of noise?

Interestingly enough, the answer to that question has probably been known for more than forty years, In 1974, the EPA's Office of Noise Abatement and Control (ONAC) (which no longer exists) conducted a detailed study on the effects of wind turbine noise on nearby properties. Although now very dated, the study still contains some of the most valuable information available on issues related to wind turbine noise, even though today's wind companies have not always made the best possible use of the study data (Ambrose and Rand 2014).

According to the ONAC study, it is relatively easy to predict the type of response humans will give to turbine noise, if one knows the level of that noise and the distance from its source. That information is summarized in a graph contained in the Ambrose and Rand article as well as in the original ONAC report itself (Ambrose and Rand 2014; "Information on Levels of Environmental Noise" 1974, Figure D7). According to the ONAC study, noise levels of less than 29 dB essentially pose no risk of objection from the general public. Among the six studies conducted with noise levels at this intensity, there were no complaints from people to governmental agencies or businesses about turbine noise. As noise level increases, however, this situation changes and a greater number of objections are lodged. For example, at about 33 dB there are sporadic complaints that can effectively be ignored by monitoring agencies. Between 33 and 40 dB, however, there begin to be a significant number of complaints with anywhere from 6 to 25 percent of the subject population raising official objections. Finally, beyond 40 dB agencies can expect to receive "strong appeals to stop the noise," and above 50 dB, members of the general population begin to organize and conduct "vigorous community action."

In reviewing the data from the ONAC study and other studies on the effects of wind turbine noise, noise control engineers Steve Ambrose and Rob Rand made the following general observation:

> Acousticians have known for decades how to predict the community reaction to a new noise source. Wind turbine consultants have chosen not to predict the community reaction as they have previously done for other community noise sources. If they had, there would be far fewer wind turbine sites with neighbors complaining loudly about excessive noise and adverse health impacts. (Ambrose and Rand 2014)

Shadow Flicker

Another effect of wind turbine operation that has been suggested as possibly responsible for human health problems is the so-called *shadow flicker* effect. As the blades on a turbine rotate, they may catch sunlight and reflect it into nearby homes, or they may alternately interrupt sunlight and cast patterns of light and shadow across nearby structures. The latter effect is called the shadow flicker effect because a person exposed to the effect sees a recurring pattern of light and darkness at an interval that corresponds to the rotation of the turbine blades. Some research suggests that this pattern may be sufficiently pronounced as to be detectable even when a person's eyes are closed (Harding, Harding, and Wilkins 2008).

The health effects of the shadow flicker effect seem to be relatively mild. These effects appear to be restricted to psychological conditions such as annoyance, stress, and sleep deprivation ("Visual Health Effects and Wind Turbines" 2014). While such problems should not be ignored, they can be avoided rather easily by appropriate planning of a wind farm siting. For example, turbine blades should not be made of a highly reflective material or, if they are, that material should be painted with a nonreflective substance. Also, wind farms should be laid out in such a way that nearby residents and businesses are not in the path of a flicker shadow ("Wind Turbine Health Impact Study" 2014, 34–38; Appendix B).

Desecration of Tribal Lands

One of the criticisms of the establishment of wind farms that is probably somewhat less often heard is the effects that turbines may have on tribal lands. The term *tribal lands* refers to areas within the United States that are largely, but not entirely, under the control of Native American tribes who occupied those lands prior to their conquest by U.S. forces. The federal government regards tribal lands as being *domestic dependent nations,* which means that they have many of the same rights and responsibilities as other sovereign nations, but that many governmental functions are still reserved to the federal government.

This somewhat ambiguous description of the rights and responsibilities of native tribes is an issue when the federal government decides to grant a permit to a company to establish a wind farm on tribal lands. Tribal members often have objections about such decisions, as they are seen to damage or destroy essential elements of tribal history and customs. For example, the Mashpee Wampanoag Tribe on Cape Cod and the Wampanoag Tribe of Gay Head/Aquinnah on Martha's Vineyard have raised objections to construction of the Cape Wind offshore wind farm at least partly because welcoming of the sunrise over Nantucket Sound has traditionally been a part of the tribes' religious ceremonies (in fact, Wampanoag means "People of the First Light") (Goodnough 2014).

In many cases, construction of a wind farm produces more concrete damage than simply the disturbance of cultural and religious traditions. Erecting dozens of very large wind turbines obviously requires extensive Earth movement and disruption of the land. Such processes may interfere with tribal burial grounds, cremation sites, paintings and geoglyphs, and other historical and sacred sites for Native American tribes (Raferty 2014). Tribes often object to wind turbines also because of the damage they cause to wildlife, especially uniquely sacred animals such as the bald eagle (a point to be discussed in more detail later in this chapter) (Zuckerman 2014).

Beyond the facts of the harm done to Native American lands and traditions is the feeling by residents of such lands that there is not much they can do to prevent the establishment of wind farms on their lands. Since the federal government has ultimate authority on nondomestic issues, of which permits for wind farms seems to be one, tribal governments can do little more than explain and emphasize the importance of damage done to their physical property and historical, religious, and cultural traditions by the construction of wind farms. As one observer has written:

> Too often, the commercial proponents and governmental permitting agencies show open disdain for the voice of tribal governments seeking to protect their few remaining cultural monuments. If they don't overtly ignore legal protections afforded to tribal cultural resources, they more often than not do the bare minimum required under federal and state law, which, in the end, is not much at all. (Sexton 2014)

It should be pointed out that the federal government has not been unaware of the issue of wind farm siting on tribal lands. Its interests, however, have been quite different from those of the tribes. For a number of years, federal authorities have made a special effort to convince tribes of the value of having wind farms on their lands, pointing out the advantages of having easy access to this "green" form of energy. The argument has been that many tribal lands are located in areas where winds blow regularly and forcefully, making them ideal sites for wind farms. Since many tribes lack access to inexpensive energy, the government has said, the creation of wind farms on tribal lands is a logical decision.

Since the end of the twentieth century, the U.S. government has created a number of aid programs to formalize this message to Native American tribes. The DOE's Wind Energy Program, for example, makes available anemometers for tribes to use in measuring wind speed, conducts workshops on the development of wind farms on tribal lands, and provides technical expertise in the planning, development, and operation of such farms. From 2003 to 2009, the DOE also published a newsletter directed specifically at tribes interested in the construction of wind farms on their lands ("Wind Power for Native Americans" 2014).

Technical Issues

A number of objections to the construction of wind farms are based on a host of technical problems, ranging from some issues of relatively minor significance to major questions that may raise questions as to whether or not wind farms should even be allowed to exist. At the less severe end of these technical issues are questions as to how the operation of wind turbines might affect communication systems in the neighborhood of a wind farm. For example, some neighbors of wind farms have reported that their television reception has been affected. They may receive less satisfactory pictures or audio signals; they may not receive some channels; or their televisions may cease to perform properly at all. As an example, residents of Langley Park, a village in Durham County, England, complained in 2008 that their television reception had been compromised after the energy company EDF Energy had installed a wind farm near the village. They either received poor reception or failed to receive some television channels at all. The company eventually acknowledged this problem and admitted that it had failed to properly test turbine function for this effect before the farm was installed. It later made adjustments that reduced or eliminated the problem, a step that it could have taken prior to beginning operation of the wind farm (Wood 2014).

More serious technical problems arise simply from the natural weather constraints associated with the generation of electricity by wind power. As noted in Chapter 1, turbine operation depends absolutely on the speed of the wind available at a wind farm location. If the wind blows too weakly, or not at all, turbines will produce little or no electricity. Generally speaking, wind velocity must reach about 10 miles per hour (4.5 meters per second), the so-called *cut-in speed,* before a turbine begins to produce any energy at all. Energy production then improves dramatically as wind velocity increases. Recall that the wind equation shows that energy production increases as the third power of wind velocity, so above 10 miles per hour, energy production begins to improve rapidly. Above a certain wind velocity, however, mechanical damage to the turbine is possible, and it must be shut down entirely. (For a graphical representation of this power function, see "Wind Turbine Power Output" 2014.) In practice, then, turbines are efficient producers of electricity when the wind blows constantly at a velocity somewhere between 10 and 60 miles per hour (27 meters per second), the so-called *cut-out speed,* but inefficiently or not at all above or below those limits.

In order to understand the problem of variable wind speeds and power production, consider a typical wind farm. Electricity produced by that wind farm varies from minute to minute, hour to hour, and day to day, depending on the speed of the wind available. Operators can attempt to predict these variations, but current tools for doing so are not very effective. As soon as a turbine begins to produce electricity, that electricity has to be transmitted away from the wind farm into the nearest electric grid. An electric grid is a network of plants that produce electrical power and consumers who use that power, connected by transmission and distribution lines and operated by one or more control centers. The grid must not only transfer power from producers to consumers *when they need it* but also do so in a smooth and orderly way. In summer, for example, consumption of electrical power increases during the day, when many air conditioners operate, and decreases at night, when the temperature falls.

This balance between energy production and consumption in a grid is possible with conventional power plants because power production at the source can fairly easily be controlled. In a power plant that operates on natural gas, for example, operators can continuously increase or decrease the amount of gas supplied to the plant and, hence, the rate of power production. In a hydroelectric plant, power production can be regulated simply by increasing or decreasing the flow of water through the plant. If the demand for electricity decreases, a plant can switch to a condition known as *spinning reserve* or *spinning standby* in which the plant continues to operate but only at a low maintenance rate. If power demand increases, operators can increase the supply of fuel, and the plant can switch over from standby reserve to *generation* to meet increased demand.

Wind farms do not have this flexibility. A turbine cannot be maintained at standby reserve if no wind is blowing; it simply stops operating completely. But when this happens, a change must be made by operators of the grid to make up for the electricity it is no longer getting from wind. Whenever a grid commits itself to accepting electricity from a wind source, it estimates the amount of electricity it can expect from that source, the *base load* from wind energy. If the amount of electricity from wind falls to less than the base load, the deficit must be made up from somewhere else, typically a traditional power plant fueled by coal, oil, natural gas, or nuclear fuel.

This situation makes it likely that the installation of a wind power facility might even *increase* the amount of fossil fuel burned in a country. Conventional power plants must continue to exist to have electrical current available when wind farms do not provide their base load. And they must be operating at least in standby mode, to make up for the loss of expected energy from the wind source.

An excess of power from a wind source can also be a problem. Imagine that wind begins to blow at a high rate of speed over an extended period of time, faster and longer than plant managers had anticipated. What happens to that excess electricity? It cannot conveniently be stored for later use. It must be transmitted to the grid, where, if demand is high, it will be used up, but, if demand is low, it must be "drained off" in some way. One of the most common ways to get rid of excess electricity is simply to ship it off to some other region: another locality, state, or even country. Denmark, with its very large wind farm installations, now ships its excess to other countries, most commonly to Norway and Sweden, when it has more than it can use at home. This system works well for all countries concerned since Denmark is able to sell its excess electricity and Norway and Sweden can store the water they use for hydroelectric power (95 percent of their electrical needs in Norway and 46 percent in Sweden) for later use while using inexpensive Danish electricity (Green, Richard 2014; Morris 2014).

An additional problem occurs when the wind increases rapidly and turbines begin to generate electricity at a rapid rate, sometimes unexpectedly. When that happens, the grid may experience a *power surge,* a sudden oversupply of electrical energy that typically lasts only a short period of time, on the order of a few microseconds. A power surge can have serious consequences in a grid because it has the potential for severely damaging devices to which electrical power is being delivered. In one example reported in 2012, a power surge in a German grid delivered so much electrical current to an aluminum plant in Hamburg that some equipment was badly damaged, and the plant had to close down briefly for repairs (Green, Kenneth 2014).

The technical problems associated with wind power are reflected in a comparison between two ways of expressing the power output of a wind turbine or wind farm. One measure is called the *rated capacity* or *nameplate capacity,* an indication of the total amount of power that can be delivered by a device or a system. Most manufacturers and proponents of wind power express the energy potential of wind systems in terms of their rated capacity, the total amount of energy they are capable of producing under ideal conditions. But no system, including wind systems, ever operates under ideal conditions. So the amount of energy actually produced by a turbine or wind farm is always some fraction of the rated capacity, a quantity known as the *capacity factor.* Capacity factors for wind systems range all the way from zero to less than about 50 percent. A number of studies have shown that the capacity factor of modern wind farms ranges between about 20 percent and 30 percent. A summary of all wind-generated power in the United States by the U.S. DOE released in January 2014, for example, indicated that the capacity factor for wind energy ranged from a low of 28.1 percent to a high of 32.3 percent in the period between 2008 and 2013 ("Capacity Factors for Utility Scale Generators" 2014, Table 6.7B). Other studies have found capacity factors for large systems even lower. A 2008 study of wind generators in Europe, for example, found that the capacity factor of 35 percent assumed by industry and governmental bodies was grossly incorrect, with the actual value closer to about 21 percent (Boccard 2009).

Wind engineers are well aware of the technological problems associated with the generation, storage, and distribution of electrical energy by wind turbines. As wind power becomes a greater factor in the energy equation of many nations, these engineers and other researchers are looking for ways to mitigate the problems caused by the irregular operation of wind turbines on the electrical grid. One example is a system developed by researchers at North Carolina State University and Johns Hopkins University in 2013. The researchers developed a system of algorithms by which excess electricity produced by wind turbines can be directed away from the grid and into a battery energy system (BES), where it is used to charge batteries. When energy generation from the wind system declines, the batteries are then discharged, releasing electrical current to the grid (Chandra, Gayme, and Chakrabortty 2013).

Another proposed solution to the variability of wind energy is the use of flywheels. A flywheel is simply a heavy wheel that spins on an axis. Adding energy to a flywheel makes it rotate more rapidly, and removing energy from the flywheel causes it to slow down. Flywheels have been used in a wide variety of mechanical devices for well over a hundred years. Today's version of the flywheel, however, is considerably more sophisticated than its ancestors, usually made of carbon fiber weighing more than 500 kilograms with a diameter of about 50 centimeters. Flywheels used as electrical backups have an angular velocity of about 30,000 revolutions per minute.

A flywheel acts as a conservation device in an electrical system by taking excess electricity produced by wind turbines to make the flywheel rotate more rapidly, converting the electrical energy into the kinetic energy of the flywheel. When demand within a grid increases, the flywheel can be slowed down, converting the reduced rotational energy into electrical energy, which can be fed into the grid. A number of research projects are currently under way, many supported by grants from the U.S. DOE, to develop flywheel systems for smoothing out the electrical current supplied by wind farms to an electrical grid (Ibrahim et al. 2011; "Solving the Use-It-or-Lose-It Wind Energy Problem" 2014; Taylor 2009; for a comprehensive technical discussion of possible solutions for the problems discussed here, see Teodorescu, Liserre, and Rodriguez 2011).

Yet another way of dealing with the irregularity of wind speed was first proposed in 1980 by American electrical engineer Miles L. Loyd in a seminal paper, "Crosswind Kite Power (for Large-Scale Wind Power Production)." Loyd suggested lifting a wind turbine into the atmosphere, some significant distance above ground, where winds tend to blow more consistently and at greater speed than they do at ground level. He predicted that such turbines could generate electricity at a significantly greater rate and more consistently than they could if built on the ground (Loyd 1980; Loyd's original paper is available online at http://homes.esat.kuleuven.be/~highwind/wp-content/uploads/2011/07/Loyd1980.pdf).

Loyd's suggestion was largely ignored for many years but has received considerably more attention in the second decade of the twenty-first century. As of late 2014, about a half-dozen companies are working on the design and/or construction of one or more variations of Loyd's so-called airborne wind turbine, also known as a high-altitude wind turbine (HAWT) (Levitan 2014).

Harm to Animals

Bird Deaths

On January 8, 2014, the American Bird Conservancy and Black Swamp Bird Observatory (ABC/BSBO) sent a letter to a number of federal governmental agencies stating their intent to sue the agencies to prevent their issuing final approval for the construction of six wind turbines on the shore of Lake Erie at the Camp Perry Air National Guard Station near Port Clinton, Ohio. ABC/BSBO argued that the turbines would pose a significant threat to eagles, migratory songbirds, and other rare and endangered species that would be killed by collision with the turbines and associated transmission lines. The government agencies responded to the ABC/BSBO letter by saying that they agreed with the complaint to the extent that they had not given adequate attention to this problem, and they also said that the project would "not go forward at this time" ("Camp Perry Wind Turbine Project" 2014; a link to the ABC/BSBO letter is found on this Web page).

Concerns about the damage caused by wind turbines to animals—most commonly birds and bats—is one of the most long-standing and contentious debates about the use of wind power. As long as wind farms have existed, environmentalists have noted the risks they pose to animals that may come into contact with the turbines themselves or with the complex transmission wires associated with the farms.

Some of the earliest—and most troubling—stories of bird kills at wind farms came from the Altamont Wind Park in California, the world's largest concentration of wind turbines. Soon after the farm began operating, naturalists noticed that large numbers of birds were being found dead on the ground within the wind park. It was obvious that the birds had been killed by flying into the rotating blades of the turbines. According to one study of the problem, about 2,700 birds are killed each year at the Altamont facility, of which 1,100 are raptors. One number that stands out from the study is the 67 golden eagles that lose their lives each year to wind turbines (Smallwood and Thelander 2008; Smallwood and Thelander 2014).

The Altamont Pass facility poses a special hazard for birds for two reasons. First, it has the largest concentration of wind turbines of any facility anywhere in the world, making it difficult for birds *not* to encounter wind machines in the area. Second, the pass is located on a major north–south migratory route taken by very large numbers of birds in their annual migrations.

Very different data on bird deaths have been reported for smaller wind facilities located in less sensitive areas. In fact, it has been difficult to get reliable date on the number of birds killed annually by wind turbines in the United States or other parts of the world. Numbers ranging anywhere from 10,000 to 573,000 deaths per year have been reported by a variety of authorities. Probably the best estimate currently available is that from a summary of studies reported by researchers from the Smithsonian Institute and the U.S. Fish and Wildlife Service in 2013. They found from a review of sixty-eight studies of bird mortalities at wind facilities that the number of birds killed annually by turbines is somewhere between 140,000 and 328,000, with a median estimate of 234,000 annual fatalities (Loss, Will, and Marra 2013; a summary of data from a number of facilities of different sizes is available at "Wind Turbine Interactions" 2014, Figures 1 and 2).

As the ABC/BSBO letter with which this section opened indicates, the greatest concern about bird deaths at the hands (or arms) of wind turbines usually focuses on eagles, other raptors, and other endangered or threatened species. In fact, by far the greatest proportion of bird deaths occurs among songbirds for a number of reasons. First, there simply *are* more songbirds than other species. Second, many songbirds undertake annual migrations that bring them into contact with wind farm facilities.

Third, those migrations often occur at night, when turbines are more difficult to see. As a consequence, about three-quarters of all bird deaths resulting from contact with wind turbines are songbirds ("Wind Turbine Interactions" 2014).

One question that often arises in discussions about bird deaths resulting from contact with wind turbines is how significant mortality rates from this cause are compared to deaths from other causes. Again, a number of studies have been conducted to develop some general estimate for these factors. One summary of those studies points out that

> the number of birds killed in wind developments is substantially lower relative to estimated annual bird casualty rates from a variety of other anthropogenic factors including vehicles, buildings and windows, power transmission lines, communication towers, toxic chemicals including pesticides, and feral and domestic cats. ("Wind Turbine Interactions" 2014)

One of the comparisons most commonly raised is the relative threat to birds from a variety of sources, such as domestic cats, building windows, and cars, all of which are responsible for a substantial number of bird deaths annually. In one review of this issue, Elliott Negin of the Union of Concerned Scientists points out that

> given how [Robert Bryce, senior fellow at the Manhattan Institute] portrays the wind industry, one would assume it's one of the nation's top bird killers. In fact, wind turbines are way down in the pecking order. (Negin 2014)

By far the greatest threat to birds in the United States, Negin says, is buildings, which account for 970 million bird deaths annually (Klem and Saenger 2013). Other risk factors for birds, Negin says, includes power lines, on which 175 million birds die each year; misapplied pesticides, accounting for 72 million bird deaths; communication towers, which kill 6.6 million birds each year; and oil and gas waste pits, responsible for as many as one million bird deaths (Negin 2014). By contrast, Negin points out, wind turbines are responsible for the death of 573,000 birds, as reported in one 2013 study (Smallwood 2013, which differs from earlier estimates by the same author). "That's not insignificant," Negin concludes, "but certainly not the scourge that Bryce implies" (Negin 2014).

Other writers have attempted to compare the cost in bird deaths of various forms of energy production. One such article, for example, points out that wind farms are responsible for approximately 0.27 bird deaths per gigawatt hour of energy produced. By comparison, nuclear power plants are responsible for 0.6 bird deaths per gigawatt hour of energy produced and fossil-fueled power plants, for 9.4 bird deaths per gigawatt hour of energy produced (Shahan et al. 2014, citing Sovacool 2014).

Bat Deaths

Birds are not the only animals killed by wind turbines. Scientists have known for many years that bats are also at high risk for collisions with wind turbines, although, as with birds, the precise number of animals killed has not been known. Estimates of bat deaths as a result of turbine collisions in the United States have ranged from a low 33,000 per year to 888,000 annually (Hayes 2013, 975). Perhaps the most reliable of recent estimates has come in a study conducted by Mark A. Hayes at the University of Colorado at Denver. Hayes studied data available on bat deaths from twenty-one wind facilities in the United States and determined that an estimated 600,000 bats were being killed annually by wind turbines (Hayes 2013). Bat deaths are of substantial importance because they are vital nocturnal predators of insects, many of which are responsible for extensive crop damage. Bats also act as pollinators for many commercial crops. One recent study estimates that the current loss of bats to wind turbines could be responsible for a loss of as much as $3.7 billion per year because of the increased cost of pest treatment occasioned by bat deaths (Boyles et al. 2011).

As with bird deaths from wind turbines, bat deaths are generally caused by bats that become confused and disoriented during their evening forays into the countryside or during migration. Death patterns vary widely from region to region and from wind farm to wind farm. The highest death rates reported by Hayes were from wind farms in Tennessee (31.5 to 53.3 deaths per megawatt per year) and the lowest from farms in Minnesota (02. deaths per megawatt per year), Oklahoma (0.5 deaths per megawatt per year), and Oregon (0.8 deaths per megawatt per year) (Hayes 2013, 976). Hayes later noted that his estimate of 600,000 bat deaths might well be conservative because of the way he conducted his study. That number could actually be as high as 900,000 bat deaths annually, he said (Flatt 2014). (Data on the number of bats killed by conventional and nuclear power plants are not currently available.)

Possible Solutions

The issue of bird and bat deaths from wind turbines is somewhat different from other problems associated wind power in that almost everyone agrees about the nature of the problem (wind turbines *do* kill bats and birds) and that something should and can be done about the problem. Even as they decry the loss of birds and bats to wind turbines, for example, most environmentalists also admit that they are strongly in favor of wind power because of its potential to reduce the world's dependence on fossil fuels. But they also say that wind companies need to work harder to reduce the loss of animal life from turbines (Schelmetic 2014).

Those wind companies have considered a variety of ways in which turbines can be made safer for birds and bats. Here are some of the approaches that have been recommended:

- *Better siting*. One of the ironies about wind farm siting is the locations that are likely to be the best locations for a wind turbine are also places that birds like to fly. For example, the wind conditions most satisfactory for a wind turbine are also the conditions preferred by eagles and other raptors. Without foresight and planning, wind companies are likely to install their turbines in just the areas over which birds are most likely to fly. The solution to this problem, then, is obvious: give more careful thought to the siting of turbines. This rule of thumb does not mean that a wind company has to abandon high-wind sites just for the sake of birds. It does suggest that careful studies of the area can reveal the specific regions that birds are most likely to be found, and turbines can then be placed in other regions that are a few hundred meters or less outside of the bird regions. Siting decisions might be no more complicated than placing a wind farm way from the edge of a cliff or a small hill, where birds are more likely to be flying (Drouin 2014; Manville 2014).
- *Radar*. Another approach that is being tried involves the use of radar systems to detect the approach of large groups of birds or bats to a wind farm. When the radar detects such an event, operation of wind turbines can be adjusted to reduce the risk posed by the turbines to the approaching animals. A company called DeTect, Inc., has developed and tested such a system with promising results. Its RAPTOR and MERLIN early detection radar systems are now installed at more than 120 locations in the United States and other countries (Davenport and Kelly 2014; see also the DeTect Web site at http://www.detect-inc.com/ for more information on the technology).
- *GPS*. A somewhat similar tracking technology is also being considered for the protection of birds from wind turbines.

One of the greatest concerns for California wind farms in particular is the risk they pose to endangered birds, such as the condor. The bird has been brought back from the brink of extinction by an ambitious conservation program, only to be faced with a new threat: wind turbines. A plan for the protection of the bird involves tracking individuals that have been fitted with radio transmitters (about half of the existing population of condors), which can be detected by global positioning system (GPS). When a wind farm detects an approaching bird, it can reduce turbine speeds or shut them down entirely. Of course, both radar and GPS technologies pose a new consideration for wind farms in addition to adding a level of protection for birds and, sometimes, bats. The irregularity of power produced by turbines because of differences in wind speeds may become even more pronounced if and when turbines also have to alter their speed because of approaching birds and bats (Sahagun 2014).

- *Ultrasound acoustics.* Protecting bats from wind turbine damage is a different problem from protecting birds. Birds die when they actually collide with turbine blades, whereas bats are killed when they get too close to the blades and their lungs literally explode because of the reduced air pressure in front of the blades (Brahic 2014), so the problem in protecting bats is to keep them from even getting close to turbine blades.
- One proposed solution for this problem has been to fit turbines with devices that send out acoustical signals that confuse a bat and make it change course away from turbine blades. This type of device has been tested and actually installed in a few wind facilities with some success (Arnett et al. 2014).
- *Turbine operation.* Another approach for bird and bat protection is to make changes in the way in which wind turbines are designed and built to reduce the likelihood of

collisions. One suggestion has been simply to reduce the wind speed at which a turbine begins to generate electricity. In most facilities, a turbine is scheduled to begin rotating when wind speeds reach 8 to 9 miles per hour. In one experiment, this standard was changed so that turbines began operating only when wind speeds reached 11 miles per hour during periods of maximum bat presence (as during migrations). Researchers found that this modest change resulted in a reduction of bat deaths anywhere from 44 to 93 percent. This change was attained even though the power output of the wind turbines involved was reduced by less than 1 percent annually (Arnett et al. 2010, 2011; see also Baerwald et al. 2009).

- *Turbine design*. Another line of research designed to make wind turbines bird- and bat-safe involves fundamental changes in the design of a turbine itself. Some inventors are looking, for example, for ways to adapt the traditional vertical-axis wind turbine for commercial use. Turbines based on the traditional Savonius or Darrieus windmill do not have the same type of rotating blades found in almost all horizontal-axis turbines in use today. They should, at least in principle, then, be safer for birds and bats. Experimental vertical-axis turbines have taken a number of quite different-looking styles, some of which can be seen at "Bird-Safe Wind Power in Alaska," "Vertical Axis Wind Turbine," and "Whole New Wind Turbine Design."

    - Other inventors and engineers have designed turbines that still contain rotating blades but that provide some type of protective shielding around the blades to prevent birds and bats from coming into contact with the turbine itself. An example is the Catching Wind Power system that claims that it is not only safe for birds and bats but also more efficient than traditional wind turbines. Many designs of this type are still in the early stages of production and may or may not ever become practical realities in the world of wind energy production. Another design has been developed by the FloDesign company, which has proposed the use of its bird- and bat-safe machines for the Altamont Pass wind farm ("FloDesign Wind Turbine" 2014).

- Yet another approach involves the painting of some part of a wind turbine a different color from the rest of the machine. Since wind turbines tend to be silvery or white, the color most often mentioned is black. The principle behind this approach is that birds are very sensitive to different colors, and it may be much easier for them to detect a wind turbine in which two different colors are displayed rather than a monochrome machine. In 2013, the Norwegian energy company Statkraft initiated an experiment in which four of the sixty-eight turbines at its Smøla site had black stripes painted on them. The experiment is designed to measure the number of birds killed by these altered wind turbines compared to those killed by monochrome turbines (Campbell 2014).

(Two very comprehensive and useful summaries of research on wind turbine–wildlife interactions and possible solutions for associated problems are "Birds and Bats Fact Sheet" 2014 and "Summary of Wind–Wildlife Interactions" 2014.)

Public Attitudes about Wind Power

The previous presentation makes it clear that many problems are still associated with the use of wind turbines as a major source of energy in the United States and most other parts of the world. Yet, it seems difficult to argue that these problems are so serious that wind power will *not* be a part of the world's energy equation in the future. Public opinion polls consistently seem to support this view. For example, a CNN/Opinion Research poll on energy conducted in March 2011 found that 83 percent of respondents said that they thought the United States should rely more on wind power in the future compared to 88 percent who named solar power, 42 percent who said nuclear power, and 28 percent who said oil ("Energy" 2014; all polling results discussed here are taken from this source).

These results were generally similar to those obtained in a Gallup Poll on energy conducted two years later in March 2013. In that poll, wind was the second most commonly mentioned energy source that should receive greater emphasis in the United States (mentioned by 71 percent of respondents) following solar power (77 percent) and outpacing oil (46 percent), nuclear power (37 percent), and coal (31 percent). When asked more specifically whether or not they would recommend a greater investment of federal funds in the development of wind and solar power, 69 percent of respondents in a 2012 Gallup poll said yes to that suggestion. Other options for dealing with the nation's energy crisis received almost the same approval ranking (higher emissions and pollution standards: 70 percent) or somewhat lower rankings (greater drilling and mining on federal lands: 65 percent and expanded use of nuclear power: 52 percent).

Global surveys of attitudes about wind energy reflect roughly similar trends as those seen in the United States, with some variation from country to country. The 2012 Global Consumer Wind Study, for example, found that about three-quarters of respondents were generally favorable toward increasing the use of wind power for energy needs in their countries, viewed companies who used wind power in the development of their products more favorably, and were somewhat more willing than not to pay more for energy generated by wind and other renewable resources than by conventional energy sources ("Global Consumer Wind Study 2012" 2012).

Arguments in Favor of Wind Power

What, then, are the most common arguments in favor of expanding the use of wind power as a way of meeting the world's energy needs?

Wind is a clean resource. The generation of electricity by wind turbines does not produce any of the pollution associated with the combustion of coal, oil, and natural gas, or the use of nuclear materials for the generation of power. One recent study has found, for example, that the use of wind power in the United States has avoided the release of 79,600 tons of nitrogen oxide and 98,400 tons of sulfur dioxide, a significant step in reducing air pollution (Schneider, Dutzik, and Sargent 2014, 5). There are also no harmful by-products involved in the generation of wind power, such as those associated with drilling, mining, and transport of the fuel from one place to another and no waste products left over to dispose of.

Wind is a renewable resource. There is no way to imagine that the world will ever run out of wind resources. While it is true that wind patterns of sufficient intensity and regularity vary from region to region, there are still untold numbers of locations at which these characteristics are sufficiently available to make wind power a real option. (For an animation showing wind movements in the United States, see http://hint.fm/wind/. Animations for other regions and the globe as a whole are also available at http://www.met.sjsu.edu/wind/streaklines.shtml and http://www.met.sjsu.edu/wind/streaklines.shtml.)

Costs are decreasing. A half century ago, the cost of generating energy with wind turbines was well beyond any amount that could justify commercial use of wind farms. Over the years, those costs have continued to decrease so that wind power is now, or will soon be, competitive with power obtained from other energy sources. That trend is especially significant as the cost of conventional (fossil) fuels continues to increase with decreasing stocks ("20 Percent Wind Energy Penetration" 2014, Figure 3–9).

Supplies are abundant. Although wind energy today makes up only a small fraction of all the energy produced in the United States and most other countries of the world, the availability of the raw material is enormous and sufficient, according to some estimates, to provide all the electrical energy needed in this and most other countries of the world (Andrews 2014).

Wind turbines are climate friendly. At a time when the world is increasingly concerned about the accumulation of carbon dioxide in the atmosphere, with consequent climate change, any energy source that has no effect on the climate—such as wind power—is bound to be popular (Sheppard 2014).

Wind power can have economic benefits. Small-scale and personal wind systems can have economic benefits to consumers. Individuals and small businesses that generate their own electricity with wind turbines may be able to sell their excess electricity to power companies. In addition, while the cost of building a wind facility can be high, those costs can often be recouped in a relatively short period of time ("What You Need to Know about Net Metering" 2014).

Wind farms can be tourist attractions. Although many complaints have been raised about the aesthetic disadvantages of large wind farms, many people consider the sight of large fields of enormous wind generators as being beautiful, and some people may be interested in visiting such sites to appreciate their beauty. As an example, the Whitelee Windfarm in Scotland, the largest wind farm in the United Kingdom, has developed a visitor center that has an exhibit of its turbines, an interactive learning center, a café, and a shop. People come not only to learn about wind turbines but also to appreciate the beauty of an industrial achievement in a natural setting ("Whitelee Windfarm Visitor Centre" 2014).

Conclusion

The contribution that wind power has made toward meeting the energy needs of the United States and many other countries of the world has increased significantly in the past half century. Proponents of the technology see only greater expansion and improvement in the future. Still, a number of technical, social, political, and economic problems remain to be solved before wind power achieves the potential that it may have as a major component of the world's energy future.

References

Ambrose, Steve, and Rob Rand. "Wind Turbine Noise Complaint Predictions Made Easy." WindWise Massachusetts. http://windwisema.org/wind-turbine-noise-complaint-predictions-made-easy/. Accessed on April 3, 2014.

Andrews, Dave. "How Much Wind Energy Is There?" Claverton Energy Research Group. http://www.claverton-energy.com/how-much-wind-energy-is-there-brian-hurley-wind-site-evaluation-ltd.html. Accessed on April 11, 2014.

Arnett, Edward B., et al. 2010. "Effectiveness of Changing Wind Turbine Cut-In Speed to Reduce Bat Fatalities at Wind Facilities." Bats and Wind Energy Cooperative and the Pennsylvania Game Commission. http://iberdrolarenewables.us/pdf/curtailment-final-report-05-15-10-v2.pdf. Accessed on April 10, 2014.

Arnett, Edward B., et al. 2011. "Altering Turbine Speed Reduces Bat Mortality at Wind-energy Facilities." *Frontiers in Ecology and the Environment.* 9(4): 209–214. http://grist.files.wordpress.com/2013/12/cut-in-et-al-2010-changing-turbine-cut-in-speed.pdf. Accessed on April 10, 2014.

Arnett, Edward B., et al. "Evaluating the Effectiveness of an Ultrasonic Acoustic Deterrent for Reducing Bat Fatalities at Wind Turbines." PLOS One. http://www.plosone.org/article/info%3Adoi%2F10.1371%2Fjournal.pone.0065794. Accessed on April 10, 2014.

"As Gas Prices Pinch, Support for Oil and Gas Production Grows." Pew Research Center for the People & the Press. http://www.people-press.org/2012/03/19/

as-gas-prices-pinch-support-for-oil-and-gas-production-grows/. Accessed on April 2, 2014.

Baerwald, Erin F., et al. 2009. "A Large-Scale Mitigation Experiment to Reduce Bat Fatalities at Wind Energy Facilities." *Journal of Wildlife Management.* 73(7): 1077–1081.

Bill Summary & Status, 95th Congress (1977–1978). H.R.5263. CRS Summary. http://thomas.loc.gov/cgi-bin/bdquery/z?d095:HR05263:@@@D&summ2=m&|TOM:/bss/d095query.html|. Accessed on March 28, 2014.

"Birds and Bats Fact Sheet." National Wind Coordinating Collaborative. http://nationalwind.org/research/publications/birds-and-bats-fact-sheet/. Accessed on April 10, 2014.

"Bird-Safe Wind Power in Alaska." National Wildlife Refuge System. http://www.fws.gov/refuges/mediatipsheet/July_2010/BirdSafeWindPowerinAlaska.html. Accessed on April 10, 2014.

Boccard, Nicolas. 2009. "Capacity Factor of Wind Power: Realized Values vs. Estimates." *Energy Policy.* 37(7): 2679–2688.

Boyles, Justin G., et al. 2011. "Economic Importance of Bats in Agriculture." *Science.* 332(6025): 41–42.

Brahic, Catherine. "Wind Turbines Make Bat Lungs Explode." New Scientist Environment. http://www.newscientist.com/article/dn14593-wind-turbines-make-bat-lungs-explode.html#.U0bVQvldV8F. Accessed on April 10, 2014.

"Camp Perry Wind Turbine Project Halted Following Threat to Sue and Petition Campaign." American Bird Conservancy. http://www.abcbirds.org/newsandreports/releases/140129.html. Accessed on April 9, 2014.

Campbell, Shaun. "Statkraft Paints It Black to Cut Bird Deaths." *Wind Power Monthly*. http://www

.windpowermonthly.com/article/1194231/statkraft-paints-black-cut-bird-deaths. Accessed on April 10, 2014.

"Camp's Tax Reform Plan Would Axe Green Energy Handouts." *The Daily Caller*. http://dailycaller.com/2014/02/27/camps-tax-reform-plan-would-axe-green-energy-handouts/. Accessed on March 29, 2014.

"Capacity Factors for Utility Scale Generators Not Primarily Using Fossil Fuels, January 2008-January 2014." *Electric Power Monthly*, January 2014. http://www.eia.gov/electricity/monthly/cpm_table_grapher.cfm?t=epmt_6_07_b. Accessed on April 7, 2014.

"Cape Wind: Project History and Overview." Bureau of Ocean Energy Management. http://www.boem.gov/Renewable-Energy-Program/Studies/Cape-Wind.aspx. Accessed on April 1, 2014.

Casey, Zoë. "Wind Farms: A Noisy Neighbor?" Renewable Energy World.com. http://www.renewableenergyworld.com/rea/news/article/2013/02/wind-farms-a-noisy-neighbor. Accessed on April 2, 2014.

Chandra, A., D. F. Gayme, and A. Chakrabortty. 2013. "Coordinating Wind Farms and Battery Management Systems for Inter-Area Oscillation Damping: A Frequency-Domain Approach." *Power Systems, IEEE Transactions*. PP(99): 1–9.

Chapman, Simon, et al. "The Pattern of Complaints about Australian Wind Farms Does Not Match the Establishment and Distribution of Turbines: Support for the Psychogenic, 'Communicated Disease' Hypothesis." PLOS One. http://www.plosone.org/article/info%3Adoi%2F10.1371%2Fjournal.pone.0076584. Accessed on April 3, 2014.

Chapter 2. Project Description. Altamont Pass Wind Resource Area. Scientific Review Committee. http://www.altamontsrc.org/alt_doc/deir/Ch02-Project%20Description_DEIR.pdf. Accessed on March 31, 2014.

Clarke, Dave. "Wind Power Problems: Alleged Problems and Objections." The Ramblings of a Bush Philosopher. http://ramblingsdc.net/Australia/WindProblems.html. Accessed on April 7, 2014.

"Coalition to Congress: End the Wind Production Tax Credit Now." Sustainable Business.com. http://www.sustainablebusiness.com/index.cfm/go/news.display/id/25240. Accessed on March 29, 2014.

Colby, W. David, et al. "Wind Turbine Sound and Health Effects: An Expert Panel Review." American Wind Energy Association and Canadian Wind Energy Association, December 2009. http://windfarmrealities.org/wfr-docs/canwea-health-dec-2009.pdf. Accessed on April 2, 2014.

"Cost and Investment Structures." From *Wind Energy—The Facts.* European Wind Energy Association, 2009. http://www.ewea.org/fileadmin/ewea_documents/documents/publications/WETF/Facts_Volume_2.pdf. Accessed on March 26, 2014.

Cummings, Jim. Wind Farm Noise 2012. The Acoustic Ecology Institute. http://www.acousticecology.org/wind/winddocs/AEI_WindFarmNoise2012.pdf. Accessed on April 2, 2014.

Daley, Beth. "The Cape Wind Battle, 2001—Present." Boston.com. http://www.boston.com/news/local/massachusetts/gallery/042810_cape_wind_timeline/. Accessed on April 1, 2014.

Damborg, Steffen. "Public Attitudes towards Wind Power." Danish Wind Industry Association. http://www.windwin.de/images/pdf/wc03041.pdf. Accessed on April 2, 2014.

Davenport, Jenny K., and T. Adam Kelly. "Using Radar-Based Mitigation to Minimize Bird and Bat Strike Risk at Wind Energy Developments." DeTect, Inc. http://nationalwind.org/wp-content/uploads/assets/research_meetings/Research_Meeting_VIII_Davenport_poster.pdf. Accessed on April 10, 2014.

Delingpole, James. "Wind Turbines ARE a Human Health Hazard: The Smoking Gun." *The Telegraph*. http://blogs.telegraph.co.uk/news/jamesdelingpole/100227983/wind-turbines-are-a-human-health-hazard-the-smoking-gun/. Accessed on April 3, 2014.

"Discussion Draft." F:\M13\CAMP\CAMP_041.XML. http://waysandmeans.house.gov/uploadedfiles/statutory_text_tax_reform_act_of_2014_discussion_draft__022614.pdf. Accessed on March 30, 2014.

Drouin, Roger. "8 Ways Wind Power Companies Are Trying to Stop Killing Birds and Bats." Mother Jones. http://www.motherjones.com/environment/2014/01/birds-bats-wind-turbines-deadly-collisions. Accessed on April 10, 2014.

"Energy." PollingReport.com. http://www.pollingreport.com/energy.htm. Accessed on April 11, 2014.

"Energy Policy Act of 1992." http://www.ferc.gov/legal/maj-ord-reg/epa.pdf. Accessed on March 28, 2014.

Enke, Brian. "Space 2014: Major Momentum Shifts Ahead?" Examiner.com. http://www.examiner.com/article/space-2014-major-momentum-shifts-ahead. Accessed on March 29, 2014.

"Environmental Impact of Wind Farms." Union of Concerned Scientists. http://www.ucsusa.org/clean_energy/our-energy-choices/renewable-energy/environmental-impacts-wind-power.html. Accessed on March 31, 2014.

Estimating U.S. Government Subsidies to Energy Sources: 2002–2008. Environmental Law Institute. September 2009. http://www.eli.org/sites/default/files/eli-pubs/d19_07.pdf. Accessed on March 30, 2014.

"The Executive Branch and National Energy Policy: Time for Renewal." Energy & Infrastructure Project, November 2012. http://bipartisanpolicy.org/sites/default/files/BPC_Governance_Report_0.pdf. Accessed on March 28, 2014.

Farboud, A., R. Crunkhorn, and A. Trinidad. 2013. "'Wind Turbine Syndrome': Fact or Fiction?" *The Journal of Laryngology and Otology.* 127(3): 222–226.

"Fast Facts about California Wind Energy." CalWEA. http://www.calwea.org/bigPicture.html. Accessed on March 31, 2014.

"Federal Production Tax Credit for Wind Energy." American Wind Energy Association. https://www.awea.org/Advocacy/Content.aspx?ItemNumber=797. Accessed on March 29, 2014.

"Financial Incentives for Renewable Energy." Database of State Incentives for Renewables and Efficiency. http://dsireusa.org/summarytables/finre.cfm. Accessed on March 29, 2014.

Flatt, Courtney. "600,000 Bats Killed at Wind Farms in 2012." Northwest Public Radio. http://nwpr.org/post/600000-bats-killed-wind-farms-2012. Accessed on April 10, 2014.

"FloDesign Wind Turbine." Avian Study Validation Project and Study. http://www.altamontsrc.org/alt_doc/p214_avian_validation_plan_1107.pdf. Accessed on April 10, 2014.

Ganapat, Priya. "Bell Labs Kills Fundamental Physics Research." Wired, August 27, 2008. http://www.wired.com/gadgetlab/2008/08/bell-labs-kills/. Accessed on March 29, 2014.

"Global Consumer Wind Study 2012." Vestas. http://www.vestas.com/~/media/vestas/media/news%20and%20announcements/pdfs/globalconsumerwindstudy2012.pdf. Accessed on April 11, 2014.

Good, Justin. 2006. "The Aesthetics of Wind Energy." *Human Ecology Review.* 13(1): 76–89.

Goodnough, Abby. "For Cape Cod Wind Farm, New Hurdle Is Spiritual." *New York Times*. http://www.nytimes

.com/2010/01/05/science/earth/05wind.html. Accessed on April 6, 2014.

Goossens, Ehren. "Siemens Says Cape Wind Farm Starts Work, Meeting Deadline." Bloomberg News. http://www.bloomberg.com/news/2013-12-16/siemens-says-cape-wind-farm-starts-work-meeting-deadline.html. Accessed on April 1, 2014.

Green, Kenneth. "Wind and Solar Power: Destabilizing the German Grid." AEIdeas. http://www.aei-ideas.org/2012/08/wind-and-solar-power-destabilizing-the-german-grid/. Accessed on April 7, 2014.

Green, Richard. "How Denmark Manages Its Wind Power." https://spiral.imperial.ac.uk/bitstream/10044/1/9852/2/How%20Denmark%20Manages%20its%20Wind%20Power%202012.pdf. Accessed on April 7, 2014.

Harding, Graham, Pamela Harding, and Arnold Wilkins. 2008. "Wind Turbines, Flicker, and Photosensitive Epilepsy: Characterizing the Flashing That May Precipitate Seizures and Optimizing Guidelines to Prevent Them." *Epilepsia*. 49(6): 1095–1098.

Hayes, Mark A. 2013. "Bats Killed in Large Numbers at United States Wind Energy Facilities." *BioScience*. 63(12): 975–979.

Hinman, Jeffry S. 2009. "The Green Economic Recovery: Wind Energy Tax Policy after Financial Crisis and the American Recovery and Reinvestment Tax Act of 2009." *Journal of Environmental Law & Litigation*. 24(1): 35–74.

Ibrahim, H., et al. 2011. "Integration of Wind Energy into Electricity Systems: Technical Challenges and Actual Solutions." *Energy Procedia*. 6: 815–824.

Indvik, Joe. "Do Wind Power Subsidies Work?" Academia.edu. https://www.academia.edu/784217/Do_Wind_Power_Subsidies_Work_Assessing_the_impact_of_

state-level_policies_on_wind_power_development_2010_ Accessed on March 29, 2014.

"Information on Levels of Environmental Noise Requisite to Protect Public Health and Welfare with an Adequate Margin of Safety." March 1974. U.S. Environmental Protection Agency. Office of Noise Abatement and Control. http://www.nonoise.org/library/levels74/levels74.htm. Accessed on April 3, 2014.

Jacobe, Dennis. "Americans Want More Emphasis on Solar, Wind, Natural Gas." Gallup Politics. http://www.gallup.com/poll/161519/americans-emphasis-solar-wind-natural-gas.aspx. Accessed on April 2, 2014

Janssen, S. A., et al. 2011. "A Comparison between Exposure-Response Relationships for Wind Turbine Annoyance and Annoyance due to Other Noise Sources." *The Journal of the Acoustical Society of America.* 130(6): 3746–3753.

Jeffrey, Roy D., Carmen M. E. Krogh, and Brett Horner. 2014. "Industrial Wind Turbines and Adverse Health Effects." *Canadian Journal of Rural Medicine.* 19(1): 21–26.

Johnsen, Erika. "Good News: Fed Govt Making Yet Another Solar 'Investment' on Our Behalf." Hot Air. http://hotair.com/archives/2012/09/29/good-news-fed-govt-making-yet-another-solar-investment-on-our-behalf/. Accessed on March 29, 2014.

Jones, Douglas N. 1980. "The National Energy Act and State Commission Regulation." *Case Western Reserve Law Review.* 30(2): 323–356. http://ipu.msu.edu/library/pdfs/jones/The%20National%20Energy%20Act-1980.pdf. Accessed on March 27, 2014.

Kennedy, Joseph. "The Sources and Uses of U.S. Science Funding." *The New Atlantis*, Summer 2012. http://www.thenewatlantis.com/publications/the-sources-and-uses-of-us-science-funding. Accessed on March 29, 2014.

Klem, Daniel Jr., and Peter G. Saenger. 2013. "Evaluating the Effectiveness of Select Visual Signals to Prevent Bird-Window Collisions." *The Wilson Journal of Ornithology.* 125(2): 406–411.

Knopper, Loren D., and Christopher A. Ollson. 2011. "Health Effects and Wind Turbines: A Review of the Literature." *Environmental Health.* 10: 78. http://www.ehjournal.net/content/10/1/78. Accessed on April 3, 2014.

Layton, Julie. "Do Wind Turbines Cause Health Problems?" How Stuff Works. http://science.howstuffworks.com/environmental/green-science/wind-turbines-health.htm. Accessed on April 3, 2014.

Levitan, Dave. "High-Altitude Wind Energy: Huge Potential—and Hurdles." Environment 360. http://e360.yale.edu/feature/high_altitude_wind_energy_huge_potential_and_hurdles/2576/. Accessed on April 19, 2014.

Logan, Jeffrey, and Stan Mark Kaplan. "Wind Power in the United States: Technology, Economic, and Policy Issues." Congressional Research Service. http://www.fas.org/sgp/crs/misc/RL34546.pdf. Accessed on March 28, 2014.

Loss, Scott R., Tom Will, and Peter P. Marra. 2013. "Estimates of Bird Collision Mortality at Wind Facilities in the Contiguous United States." *Biological Conservation.* 168: 201–209.

Loyd, M. L. 1980. "Crosswind Kite Power (for Large-Scale Wind Power Production)." *Journal of Energy.* 4(3): 106–111.

Madrigal, Alexis C. "The 8 Biggest Wind Farms in the World." *The Atlantic.* http://www.theatlantic.com/technology/archive/2011/11/the-8-biggest-wind-farms-in-the-world/249174/#slide1. Accessed on March 31, 2014.

"Maine: Incentives/Policies for Renewables and Efficiency." DSIRE. http://dsireusa.org/incentives/incentive.cfm?Incentive_Code=ME13F&re=0&ee=0. Accessed on March 28, 2014.

Manville, Albert M., II. "Towers, Turbines, Power Lines, and Buildings—Steps Being Taken by the U.S. Fish and Wildlife Service to Avoid or Minimize Take of Migratory Birds at These Structures." *Proceedings of the Fourth International Partners in Flight Conference: Tundra to Tropics,* 262–272. http://www.partnersinflight.org/pubs/mcallenproc/articles/pif09_anthropogenic%20impacts/manville_pif09.pdf. Accessed on April 10, 2014.

"Massachusetts: Incentives/Policies for Renewables & Efficiency." DSIRE. http://dsireusa.org/incentives/index.cfm?re=0&ee=0&spv=0&st=0&srp=1&state=MA. Accessed on March 29, 2014.

McVeigh, James, et al. 1999. "Winner, Loser, or Innocent Victim? Has Renewable Energy Performed As Expected?" Washington, DC: Resources for the Future. http://www.rff.org/Documents/RFF-DP-99-28.pdf. Accessed on March 19, 2014.

"Measuring Wind Turbine Noise." Renewable Energy World.com. http://www.renewableenergyworld.com/rea/news/article/2010/11/measuring-wind-turbine-noise. Accessed on April 2, 2014.

"Mission Statement." American Energy Alliance. http://americanenergyalliance.org/about/mission-statement/. Accessed on March 29, 2014.

Morris, Craig. "Denmark Surpasses 100 Percent Wind Power." Energy Transition: The German Energiewende. http://energytransition.de/2013/11/denmark-surpasses-100-percent-wind-power/. Accessed on April 7, 2014.

National Research Council. Committee on Environmental Impacts of Wind-Energy Projects. 2007. *Environmental Impacts of Wind-Energy Projects.* Washington, DC: National Academies Press.

Negin, Elliott. "Wind Energy Threat to Birds Is Overblown." *HuffPost Green.* http://www.huffingtonpost.com/

elliott-negin/wind-energy-threat-to-bir_b_4321113.html? Accessed on April 9, 2014.

Norberg-Bohm, Vicki. 2000. "Creating Incentives for Environmentally Enhancing Technological Change: Lessons from 30 Years of U.S. Energy Technology Policy." *Technological Forecasting and Social Change* 65(2): 125–148.

"Offshore Center Danmark." Project POWER. http://www.offshore-power.net/Files/Filer/danish_supply_chain_study.pdf. Accessed on March 31, 2014.

"Optimizing Large Wind Farms." *Science Daily*. http://www.sciencedaily.com/releases/2010/11/101123174322.htm. Accessed on March 31, 2014.

Pedersen, Eja, et al. 2009. "Response to Noise from Modern Wind Farms in the Netherlands." *The Journal of the Acoustical Society of America.* 126(2): 634–643.

Philipkoski, Kristen. "Celera Wins Genome Race." Wired. http://www.wired.com/science/discoveries/news/2000/04/35479. Accessed on March 29, 2014.

Pierpont, Nina. 2009. *Wind Turbine Syndrome: A Report on a Natural Experiment.* Santa Fe, NM: K-Selected Books.

"Production Tax Credit for Renewable Energy." Union of Concerned Scientists. http://www.ucsusa.org/clean_energy/smart-energy-solutions/increase-renewables/production-tax-credit-for.html. Accessed on March 29, 2014.

"The Public Utility Regulatory Policies Act." American History. Powering the Past. http://americanhistory.si.edu/powering/past/history4.htm. Accessed on March 27, 2014.

"Public Utility Regulatory Policies Act of 1978." U.S. House of Representatives. http://www.house.gov/legcoun/Comps/PURPA78.PDF. Accessed on March 27, 2014.

"Public Utility Regulatory Policy Act." Union of Concerned Scientists. http://www.ucsusa.org/clean_energy/

smart-energy-solutions/strengthen-policy/public-util ity-regulatory.html. Accessed on March 28, 2014.

Putnam, Palmer Cosslett. 1948. *Power from the Wind.* New York: Van Nostrand Reinhold.

Raferty, Miriam. "Eight Tribal Nations Mourn Losses at Ocotillo Wind Site." *East County Magazine*. http://eastcounty magazine.org/node/10162. Accessed on April 6, 2014.

Ragheb, M. "Historical Wind Generators Machines." http://mragheb.com/NPRE%20475%20Wind%20Power%20 Systems/Historical%20Wind%20Generators%20Machines .pdf. Accessed on March 26, 2014.

Ransom, John. "Going Green Is Gauche on Robert Kennedy's Private Ocean." Townhall Finance.com. http://finance.townhall.com/columnists/johnransom/2012/06/23/going_green_is_gauche_on_robert_kennedys_vineyard/page/full. Accessed on April 1, 2014.

"Renewable Electricity Production Tax Credit (PTC)." Energy.gov. http://energy.gov/savings/renewable-electricity-production-tax-credit-ptc. Accessed on March 28, 2014.

"Renewable Energy Production Incentive." Department of Energy. http://apps1.eere.energy.gov/repi/. Accessed on March 28, 2014.

"Rep. Paul Broun Says Innovation Would Flower If Government 'Would Get Out of the Way'." *Tampa Bay Times* PolitiFact.com. http://www.politifact.com/truth-o-me ter/statements/2011/jan/31/paul-broun/rep-paul-broun-says-innovation-would-flower-if-gov/. Accessed on March 29, 2014.

Richardson, Peter, and Ken Kaufmann. "Utilizing PURPA in Today's Deregulated Wholesale Market." http://lklaw.com/wordpress_dev2/wp-content/uploads/2012/08/5June2012-Utilizing-PURPA-in-to days-Deregulated-Wholesale-Market.pdf. Accessed on March 28, 2014.

Righter, Robert W. 1996. *Wind Energy in America: A History.* Norman: University of Oklahoma Press.

Sahagun, Louis. "Terra-Gen Gets OK on Wind Farm in Wake of Condor Decision." *Los Angeles Times.* http://articles.latimes.com/2013/may/24/local/la-me-0525-condor-permit-20130525. Accessed on April 10, 2014.

Salt, Alec N. "Wind Turbines Can Be Hazardous to Human Health." http://oto2.wustl.edu/cochlea/wind.html. Accessed on April 2, 2014.

Schelmetic, Tracey. "An Environmentalist's Dilemma: Birds or Wind Turbines?" Thomasnet. http://news.thomasnet.com/IMT/2011/09/07/an-environmentalists-dilemma-birds-or-wind-turbines/. Accessed on April 10, 2014.

Schneider, Jordan, Tony Dutzik, and Rob Sargent. "Wind Energy for a Cleaner America II." Environment California Research Policy Center. http://environmentcaliforniacenter.org/sites/environment/files/reports/CA_WindEnergy_scrn.pdf. Accessed on April 11, 2014.

Schweiger, Larry, and Tom Kiernan. "Clean Energy Needs Congress's Support, Not Its Cold Shoulder." The Hill. http://thehill.com/opinion/op-ed/196722-clean-energy-needs-congresss-support-not-its-cold-shoulder. Accessed on March 29, 2014.

Seelye, Katharine Q. "Big Wind Farm Off Cape Cod Gets Approval." *New York Times.* http://www.nytimes.com/2010/04/29/science/earth/29wind.html?pagewanted=all&_r=0. Accessed on April 1, 2014.

"7 Future Wind Farms." *HuffPost Green.* http://www.huffingtonpost.com/2010/05/26/7-future-wind-farms-photo_n_588793.html#s93231&title=Titan_Wind_Project. Accessed on March 31, 2014.

Sexton, Joe. "Sacred Indian Sites Are Desecrated While Congress Fiddles." Indian Country. http://indiancoun

trytodaymedianetwork.com/2013/11/13/sacred-indian-sites-are-desecrated-while-congress-fiddles. Accessed on April 6, 2014.

Shahan, Zachary, et al. "Wind Farm Bird Deaths vs Fossil Fuel & Nuclear Power Bird Deaths." *Clean Technica.* http://cleantechnica.com/2013/11/26/wind-farm-bird-deaths-fossil-fuel-nuclear-bird-deaths/. Accessed on April 9, 2014.

Sheppard, Kate. "Wind Power Has Cut U.S. Carbon Dioxide Emissions by 4.4 Percent." *HuffPost Green.* http://www.huffingtonpost.com/2014/04/04/wind-power-emissions_n_5087308.html. Accessed on April 11, 2014.

Sherlock, Molly F. "Energy Tax Policy: Historical Perspectives on and Current Status of Energy Tax Expenditures." Congressional Research Service. May 2, 2011. http://www.leahy.senate.gov/imo/media/doc/R41227EnergyLegReport.pdf. Accessed on March 28, 2014.

Singer, Peter. "Federally Supported Innovation: 22 Examples of Major Technology Advances That Stem from Federal Research Support." MIT Washington Office, January 2014. http://www.sciencecoalition.org/downloads/1390490336mitpetersingerfederallysupportedinnovationswhitepaperjan2014-21.pdf. Accessed on March 29, 2014.

Smallwood, K. S. 2013. "Comparing Bird and Bat Fatality-Rate Estimates among North American Wind-Energy Projects." *Wildlife Society Bulletin.* 37(1): 19–33.

Smallwood, K. S., and Carl Thelander. 2008. "Bird Mortality in the Altamont Pass Wind Resource Area, California." *The Journal of Wildlife Management.* 72 (1): 215–223.

Smallwood, K. S., and C. G. Thelander. "Bird Mortality at the Altamont Pass Wind Resource Area." National Renewable Energy Laboratory. http://www.biologicaldiversity.org/campaigns/protecting_birds_of_prey_at_altamont_pass/pdfs/

Smallwood_and_Thelander_2005.pdf. Accessed on April 7, 2014.

Smith, Eric R. A. N., and Holly Klick. "Explaining NIMBY Opposition to Wind Power."

http://www.polsci.ucsb.edu/faculty/smith/wind.pdf. Accessed on April 2, 2014.

"Smith-Putnam Industrial Photos." Paul Gipe. WindWorks.org. http://www.wind-works.org/cms/index.php?id=223. Accessed on March 26, 2014.

"Solving the Use-It-or-Lose-It Wind Energy Problem." Windpower Engineering Development. http://www.windpowerengineering.com/design/electrical/power-storage/solving-the-use-it-or-lose-it-wind-energy-problem/. Accessed on April 7, 2014.

Sovacool, Benjamin K. 2012. "The Avian and Wildlife Costs of Fossil Fuels and Nuclear Power." *Journal of Integrative Environmental Sciences.* 9(4): 255–278.

Summary of Hearings on Energy. American Geosciences Institute. http://www.agiweb.org/gap/legis112/energy_hearings.html. Accessed on March 9, 2014.

"Summary of Wind–Wildlife Interactions." American Wind Wildlife Institute. http://awwi.org/resources/summary-of-wind-wildlife-interactions/. Accessed on April 10, 2014.

Taylor, Phil. "Can Wind Power Be Stored?" *Scientific American.* September 28, 2009. http://www.scientificamerican.com/article/wind-power-turbine-storage-electricity-appliances/. Accessed on April 7, 2014.

Teodorescu, Remus, Marco Liserre, and Pedro Rodriguez. 2011. *Grid Converters for Photovoltaic and Wind Power Systems.* New York: Wiley-IEEE Press.

"Tourism Blown Off Course by Turbines." *Berwickshire News.* http://www.berwickshirenews.co.uk/news/local/all-news/

tourism-blown-off-course-by-turbines-1-2859768. Accessed on April 2, 2014.

"20 Percent Wind Energy Penetration in the United States: A Technical Analysis of the Energy Resource." Black & Veatch Report for the American Wind Energy Association. http://www.20percentwind.org/Black_Veatch_20_Percent_Report.pdf. Accessed on April 11, 2014.

"2012 Annual Report." World Wind Energy Association. http://www.wwindea.org/webimages/WorldWindEnergyReport2012_final.pdf. Accessed on April 1, 2014.

Vermeulen, H. 1974. "The Economics of Using Wind Power for Electricity Supply in the Netherlands and for Water-supply on Curacao." NASA Translation. http://archive.org/stream/nasa_techdoc_19750002515/19750002515_djvu.txt. Accessed on March 26, 2014.

"Vertical Axis Wind Turbine." http://www.seao2.com/vawt/#vawts. Accessed on April 10, 2014.

"The View." Save Our Sound. http://www.saveoursound.org/cape_wind_threats/view/. Accessed on April 1, 2014.

"Visual Health Effects and Wind Turbines." The Society for Wind Vigilance. http://www.windvigilance.com/about-adverse-health-effects/visual-health-effects-and-wind-turbines#_edn10. Accessed on April 3, 2014.

"What You Need to Know about Net Metering." Consumers Energy. http://www.consumersenergy.com/uploadedFiles/CEWEB/SHARED/Customer_Generation/Net-Metering.pdf. Accessed on April 11, 2014.

"Whitelee Windfarm Visitor Centre." Scottish Power Renewables. http://www.whiteleewindfarm.co.uk/visitor_centre?mainimage. Accessed on April 11, 2014.

"Whole New Wind Turbine Design." New Energy News. http://newenergynews.blogspot.com/2010/10/quick-news-10-19-charge-for-bev.html. Accessed on April 10, 2014.

"Who's Behind the Smears?" Fight Clean Energy Smears . http://fightcleanenergysmears.org/behind_the_smears .cfm. Accessed on March 30, 2014.

"Why Wind Energy?" Windustry. http://www.windustry.org/wind-basics/why-wind-energy. Accessed on March 29, 2014.

"Wind Energy Conversion Systems." file:///C:/Users/David%20Newton/Downloads/9781848000797-c1.pdf. Accessed on April 6, 2014.

"Wind Farms and Health—2014." Australian Medical Association. https://ama.com.au/position-statement/wind-farms-and-health-2014. Accessed on April 2, 2014.

"Wind Power for Native Americans." U.S. Department of Energy. http://www.windpoweringamerica.gov/nativeam ericans/. Accessed on April 6, 2014.

"Wind Turbine Health Impact Study: Report of Independent Expert Panel." January 2012. http://www.mass.gov/eea/docs/dep/energy/wind/turbine-impact-study.pdf. Accessed on April 3, 2014.

"Wind Turbine Interactions with Birds, Bats, and Their Habitats: A Summary of Research Results and Priority Questions." National Wind Coordinating Collaborative. http://www1.eere.energy.gov/wind/pdfs/birds_and_bats_fact_sheet.pdf. Accessed on April 9, 2014.

"Wind Turbine Power Output Variation with Steady Wind Speed." Windpower Program. http://www.wind-power-program.com/turbine_characteristics.htm. Accessed on April 6, 2014.

"Wolds Wind Farm Opposition." http://www.wwfo.co.uk/ak_articles/how_much_space.html. Accessed on March 31, 2014.

Wolsink, Maarten. 2007. "Wind Power Implementation: The Nature of Public Attitudes: Equity and Fairness instead of 'Backyard Motives'." *Renewable and Sustainable Energy Reviews.* 11(6): 1188–1207.

Wood, Kerry. "Company Admits Wind Farm Ruined TV Reception." Chronicle Live. http://www.chroniclelive.co.uk/news/north-east-news/company-admits-wind-farms-ruined-1477000. Accessed on April 6, 2014.

Zuckerman, Laura. "Native Americans Decry Eagle Deaths Tied to Wind Farms." *Reuters.* http://www.reuters.com/article/2013/06/13/us-usa-eagles-wind-idUSBRE95C1GH20130613. Accessed on April 6, 2014.

3 Perspectives

Introduction

A wide range of individuals around the world have a variety of specialized interests in the topic of wind energy. Some people feel very strongly in favor of or opposed to the development and use of wind power and write and act to promote their positions. Some of the essays in this chapter take one or the other of these positions. Other individuals are generally supportive of wind energy and are especially interested in specific aspects of the topic, such as the development of wind power for small facilities or research on above-ground wind facilities. This chapter also includes essays along these lines. In all cases, contributors are representing their own views on the topic of wind power and not necessarily those of the author or the publisher.

AGAINST THE WIND: WIND POWER OPPOSITION IN AUSTRALIA

Neil Barrett

The Australian wind industry has in recent years encountered strong opposition from a loose alliance of climate change deniers, conservative media commentators, think tanks and politicians, wealthy rural landholders, and ordinary citizens

---

The Altamont Pass wind farm in California is composed of over 6,000 relatively small wind turbines of various types, making it, at one time, the largest wind farm in the world in terms of capacity. (Terrance Emerson/Dreamstime.com)

simply worried about change. The strength of this opposition has come as a great surprise to people in the wind industry and their supporters, especially those trying desperately to do something about Australia's high level of carbon dioxide (CO2) emissions per capita.

The complaints about wind power in Australia were, until 2009, much the same all over the world and related to property values, visual impact, noise, resentment of outsiders, and, to a minor extent, health concerns, as well as deeper underlying reasons such as opposition to green political agendas and the feeling in many rural communities that urban people get a better deal than they do (Acceptance of Rural Wind Farms in Australia—A Snapshot).

However, following the publication of Nina Pierpont's book *Wind Turbine Syndrome,* in 2009, it appears that—with the indispensable aid of Google, email, and social media—health concerns quickly went to the top of the list in Australia and North America and to a lesser extent, the United Kingdom[1]. No longer was there simply a perceived relationship between excessive noise- and stress-related illness: people started to actually believe that infrasound generated by turbines had direct and serious pathological impacts on the body.

The Nocebo Effect

Today it is common to hear of claims about potential adverse health effects being made prior to turbines being built. When they start operating, the same people then often claim actual effects. Simon Chapman has made a study of all fifty-one Australian wind farms operating in 2012. He concluded that this so-called nocebo effect (a negative effect caused by the suggestion or belief that something is harmful) is alive and well as (1) roughly half of all large wind farms have had no complaints, suggesting that exogenous factors to the turbines (e.g., a strong opposition group) may explain the presence or absence of complaints, (2) the total number of complaints within 5 kilometers of turbines was only 120 out of a population of almost 33,000, thereby suggesting that individual or social factors may be at work rather than health issues that should affect a greater number of people, and (3) although 70 percent of the wind farms began operating prior to 2009, 82 percent of the complaints were received after that date, suggesting that the raising of health issues in that year had a decided impact.

Chapman's conclusions are supported by work carried out at by Fiona Crichton at the University of Auckland (Crichton et al. 2013). In a peer-reviewed article published in 2013, she reported that when subjects who had been told that infrasound was very hazardous to health were exposed to sham infrasound, they still reported adverse symptoms. Those subjects who had been told that expert studies showed that infrasound was not harmful at all reported no ill effects.

While these studies indicate that the nocebo effect may be playing a significant role in causing opposition, further study of each wind farm will be necessary before we can tell how significant it is compared with those factors already mentioned.

The Groups: Landscape Guardians and the Waubra Foundation

In recent years, most new wind farm proposals have been opposed by both local and national protest groups. Members of newly formed local groups, often led by wealthy landowners concerned about the views from their properties and the possibility that property values may decline, prepare submissions to local and state planning authorities, usually with the help of parent and sister groups and, of course, the Internet, where one-sided "research" has never been easier. Typically, a local Landscape Guardians group will be set up and obtain advice, sometimes from its parent, the Australian Landscape Guardians (ALG), but more commonly these days from the ALG's close relative, the Waubra Foundation.

Not only was 2009 the year of publication of Pierpont's book in the United States and other English-speaking countries; it was also the year the 128-turbine, Waubra wind farm commenced operation. Not long after, in February 2010, the Waubra Disease Foundation was established. A few months later, perhaps on the advice of a public relations company, it dropped the word *disease* from the name. The Foundation's broad aim is to gather and disseminate information on the wind industry's impact on public health and to provide advice to individuals and communities. Although it is obliged by its constitution to maintain complete independence from government, industry, and advocacy groups for or against wind turbines, in practice it is firmly enmeshed in anti-wind and, in the case of its founder and chairman, Peter Mitchell, pro-fossil fuel networks. Mitchell was the founder of Moonie Oil Ltd. and maintains involvement in fossil fuel investments via the company Lowell Resources, which until recently shared a postal address with the Foundation. Furthermore, most of the Foundation's board members have been associated with at least one Landscape Guardians group and have attempted to fight off wind farm developments near their rural retreats.

The Foundation is now widely seen as the nation's leading anti-wind organization and its CEO, Sarah Laurie, an unregistered medical doctor, as the opposition's leading spokesperson. It is strongly represented at most, if not all, inquiries into wind farms and is frequently invited to have a representative at public meetings wherever a new wind farm is proposed. Pro-wind groups view it as a principal cause of the anxiety, stress, and ill-health it claims are a result of wind farms.

Lips Trembling at 10 Kilometers (7 Miles)

Certainly Laurie is prone to making incredible statements that test the credulity of her listeners. For example:

> Some people [living near turbines] are finding that they need to get up to go to the toilet a lot more at night, again

> correlating to specific wind directions. There's stories of places, and in one house in particular in one location, where it's a seaside location and there were lots of people staying, just about everybody was up on one particular night every five or ten minutes needing to go to the toilet. (Chapman 2014).

She has also told meetings that people's lips may tremble 10 kilometers away from turbines ("various people have described symptoms where they have described either chest or lip vibration, the lip vibrations have been described to me as from a distance of 10 kilometers away") (Barnard 2014) and suggested that at a distance of 1 kilometer, turbines might actually rock a stationary car (Waubra Foundation 2014).

Not surprisingly, Laurie's statements have often been resoundingly criticized by experts in the field, such as Adelaide University's professor of medicine, Gary Wittert (Barnard 2014) and one of the world's leading acousticians, Geoffrey Leventhall. Leventhall, an expert witness at the Senate inquiry, claimed that Laurie persistently misrepresents his work and that "she is a person who only believes what she wishes to believe and will either reject new information or bend it to support existing beliefs" (Leventhall 2014).

But severe criticism hasn't deterred Laurie. In mid-2011, the Waubra Foundation sent what it called an Explicit Cautionary Notice to relevant government bodies and private companies associated with wind farm development. After listing the alleged health problems caused by turbines and claiming that twenty Australian families—an exaggerated figure sometimes raised to forty—have had to leave their homes as a result, it concluded with this warning, which is now being used by anti-wind farm groups in Europe and North America:

> We remind those in positions of responsibility for the engineering, investment and planning decisions about

> project and turbine siting, that their primary responsibility is to ensure that developments cause no harm to adjacent residents; and, if there is possibility of any such harm, then the project should be re-engineered or cancelled. To ignore existing evidence by continuing the current practice of siting turbines close to homes is to run the dangerous risk of breaching a fundamental duty of care, thus attracting grave liability. ("Explicit Cautionary Note" 2014)

Late in 2013, a large number of Waubra people finally decided they'd had enough. A petition signed by 250 individuals in and around the town of 500 people was organized for presentation to the Foundation at its Melbourne office (McGrath 2014). Optimists saw this as a sign that ordinary Australians might be starting to fight back against the anti-wind movement.

Finally, a brief word about financing: Although there is very big money available at the higher levels of the movement, it is very likely that local groups are funded from small donations from within their own circles. Apart from those in the Institute of Public Affairs (IPA), the IPA Astro-turfed offshoot, the Australian Environment Foundation (AEF), the anonymous Web site Stop These Things, and the Waubra Foundation administration, there appear to be no paid full-time workers involved with wind farm opposition groups. In 2010, as noted earlier, the AEF organized a major demonstration on behalf of a local group at the opening ceremony of a community wind farm (Hepburn Wind 2014), but that appears to have been a one-off event.

Today (May 2014), the highly conservative, ideologically driven coalition government that came into power in September 2013 is winding back nearly all the progress Australia has made in renewable energy and other action on climate change. Because of doubt about its future, the wind industry is planning few, if any, new developments, and progress has come to a virtual halt.

Sadly, Australia risks becoming a laughing stock, if not a pariah, in a world that increasingly faces a climate change crisis.

References

Acceptance of Rural Wind Farms in Australia—A Snap shot. CSIRO. file:///C:/Users/David%20Newton/Down loads/Acceptance%20of%20rural%20wind%20farms%20 in%20Australia%20a%20snapshot_CSIRO2012_Sum mary.pdf. Accessed on May 29, 2014.

Barnard, Mike. "Bad Day in Court for Anti-Wind Campaigner Sarah Laurie." REnew Economy. http://renew economy.com.au/2012/bad-day-in-court-for-anti-wind-campaigner-sarah-laurie. Accessed on May 29, 2014.

Chapman, Simon. "Wind Turbines Power Mass Hysteria." Friends of the Earth Australia. http://www.foe.org.au/ wind-turbines-power-mass-hysteria. Accessed on May 29, 2014.

Crichton, Fiona, et al. 2013. "Can Expectations Produce Symptoms from Infrasound Associated with Wind Turbines?" *Health Psychology*. 33(4): 360–364.

"Explicit Cautionary Note." Waubra Foundation. http://wau brafoundation.org.au/about/explicit-cautionary-notice/. Accessed on May 29, 2014.

Hepburn Wind. http://hepburnwind.com.au/downloads/2010 1012%20HW%20Enews.pdf. Accessed on May 29, 2014.

Leventhall, Geoff (letter). http://webcache.googleusercon tent.com/search?q=ache:tlRVsJ1vjmMJ:https://senate.aph .gov.au/submissions/committees/viewdocument.aspx%3 Fid%3D93e4908b-d2a9-4cca-8f60-a27d99c32fe2+&cd=1 &hl=en&ct=clnk&gl=au. Accessed on May 29, 2014.

McGrath, Gav. "Waubra Wind Farm: Locals Petition to Save Town's Name." *The Courier*. http://www.thecour ier.com.au/story/1889423/

waubra-wind-farm-locals-petition-to-save-towns-name/. Accessed on May 29, 2014.

Waubra Foundation (letter). http://docs.wind-watch.org/Laurie-Collector.pdf. Accessed on May 29, 2014.

Neil Barrett was the founder and CEO of Video Education Australasia, a leading supplier of videos and DVDs to educational institutions in Australia and overseas. Since selling the business in 2001, he has been involved in climate change research and action. In 2013 he produced a series on videos on the Waubra wind farm in SE Australia and contributed a chapter for a forthcoming book on the global wind power opposition, to be published by the World Wind Energy Association.

WIND ENERGY: HOW WE GOT HERE

John Droz Jr.

The first practical use of electricity (in the late 1800s) is generally attributed to Thomas Edison (a founder of GE). Of course there were actually dozens of other people who contributed to making commercial electricity a reality—and there were *many* formidable hurdles to overcome.

One of the initial primary issues was: where was this electricity going to come from? For the first hundred years or so, there were six overriding concerns about commercial electricity generators. Could they:

1 provide large amounts of electricity?
2 provide reliable and predictable electricity?
3 provide dispatchable electricity?
4 service one or more grid demand element?
5 have a compact facility?
6 provide economical electricity?

(A dispatchable source generates power on a human-defined schedule. Grid demand element is equal to the base load plus load following [power output responds to moment-to-moment changes in system demand] plus peak load [the maximum load during a specified time period]). Compact is the ability to site an electrical facility on a relatively small and well-defined footprint, preferably near high demand, such as cities. This saves on expensive transmission lines, which can have significant power loss.

A primary goal of these efforts was to achieve capacity. To ensure reliability at the lowest cost, grid operators consider capacity in several ways as they evaluate electricity sources—but the most important is capacity value. The layperson's definition of this term is: "the percentage of a source's rated capacity that grid operators can be confidant will be available during future times of greatest demand." Knowing this accurately is the key to reliable system grid performance!

Back to our history: several options were proposed to satisfy the previously mentioned criteria. To maximize public benefit, each was individually and scientifically vetted to ascertain whether the suggested source would comply with *all* of the needed conditions.

Our careful implementation of these factors has resulted in the world's most successful grid system.

Over time, what resulted from these assessments was that we selected the following sources to provide commercial electricity: hydroelectric, coal, nuclear, natural gas, and oil. (Oil is the smallest source, supplying only about 1 percent of electricity generated in the United States.)

Note that each of these current sources meets *all* of the six essential criteria mentioned earlier—and if they don't, then they get replaced—by conventional sources that do meet all criteria.

As a result, today, and a hundred years from now, these sources can provide *all* of the electrical needs of our society—and continue to meet all six criteria.

Note that *all* of the primary conventional sources use home-grown energy. Regarding our electrical energy sources, we have always been energy independent!

So what's the problem?

Ahhh, the problem is that a new element has been recently added to the list of requirements: environmental impact. The current number-one environmental impact consideration is greenhouse gas emissions such as carbon dioxide.

Why has this joined the Big Six? It is a direct result of the current debate on global warming. Despite what the media conveys, this is not yet a scientifically resolved matter. In response to intense political pressure, our government has acquiesced to making emissions an additional criterion.

Having the government mandate that utility companies change the principles that have been the foundation of our electrical supply system for a hundred years—for reasons not yet scientifically resolved—is rather disconcerting.

And there's more. Concern #3 is that this new standard for electrical supply sources now has taken priority over *all the other six*! In fact, this new-boy-on-the-block has in reality become the *only* important benchmark—the other six are now given only lip service!

In this unraveling of sensibility, there is one final incredible insult to science: commercial electricity alternative sources that claim to make a consequential impact on carbon dioxide don't even have to prove that they actually do it!

Let's look at the environmental poster child—wind energy—and examine these criteria:

1—Does industrial wind energy provide large amounts of electricity?

Yes, it could. However, its effectiveness from most perspectives is inferior. For instance (because of the wide fluctuations of wind), on average, it produces only about 30 percent of its nameplate power. Then compare these energy densities (MJ/kg): nuclear=88,000,000, gas=46, wind=.00006.

2—Does industrial wind energy provide reliable and predictable electricity?

No. Despite the wind industry's best efforts, it is not reliable or predictable compared to the standards set by our conventional electrical sources. What's worse is that when power is really needed (e.g., hot summer afternoons), wind is usually on vacation.

3—Does industrial wind energy provide dispatchable electricity?

No. Due to its unpredictability, wind can't be counted on to provide power on-demand, that is, on a human-defined schedule.

4—Does industrial wind energy provide one or more of the grid demand elements?

No. It cannot provide base load power, load following, or peak load. Essentially, wind energy is just thrown into the mix and gets used who knows where.

5—Is industrial wind energy compact?

No. To even approximate the nameplate power of a conventional facility, like nuclear, takes over a thousand times the amount of area. Wind promoters try to convince nontechnical politicians that it can have real capacity value. Their Tinkertoy "solution" is to connect multiple wind farms spread over vast areas (often several states). In Australia it has been proven that this doesn't work. Even if it did, this would undermine the objective to be a concentrated power source.

Another "feature" of wind energy is that most of the windiest sites (and available land) are a *long* way from where the electricity is needed. This will result in thousands of miles of transmission towers and cables, at an enormous expense to ratepayers.

6—Does industrial wind energy provide economical electricity?

No. It is artificially subsidized *way* more than any conventional power source. A 2008 report by the U.S. Energy Information Administration concluded that wind energy is subsidized to the tune of $23 per megawatt-hour. By contrast, normal coal receives 44¢ per megawatt-hour, natural gas 25¢, hydroelectric 67¢, and nuclear power $1.59 (Since these other sources meet *all* six criteria, there is some basis for subsidizing them!) ("Direct Federal Financial Interventions and Subsidies" 2014).

And now the latest rule du jour:

7—Does industrial wind energy make a consequential reduction of carbon dioxide?

No! No scientific study has ever proven that wind energy saves a meaningful amount of carbon dioxide. In fact, the most scientific study done (by the National Academies of Sciences) says the opposite. This 2007 report concludes that (assuming the most optimistic conditions) the U.S. carbon dioxide savings by 2020 will amount to only 1.8 percent. This is a trivial quantity and amounts to about 1/80,000 of the world's carbon dioxide emissions ("Environmental Impacts of Wind-Energy Projects 2014).

What about the critical factor of capacity value? The result of the previously mentioned deficiencies is that wind energy has a capacity value of about zero. Compare this to the conventional sources, where essentially all of them have a capacity value near 100 percent—a stunning disparity.

How can this possibly be? How could the United States be on the path to spend over a *trillion* dollars on an electrical source that fails five out of six of our historically important criteria *and* has no scientific proof that it even meets this new emissions criterion?

It's all about the money. Lobbyists for parties who want a piece of this *trillion* dollars are leaving no stone unturned. Environmentalists who have taken their eye off the ball are promoting this palliative non-solution. Politicians eager to be seen as "green" are saying yes to everything the color of money.

Wind energy proponents typically try to rationalize away its serious shortcomings, saying that things will "get worked out" mañana. What essentially is happening though is that our politicians are trying to pound a square peg into a round hole. Zero wind energy is appropriate until after these significant problems are resolved—as some may never be (due to the laws of physics).

After understanding wind energy's inherent electrical generation defects, it might put some other issues into perspective.

For instance, it is entirely legitimate to be concerned about bird and bat mortality, noise intrusions, property devaluation, and so on. But what if they were "fixed" (with a protective ordinance)—would wind energy then be OK?

No: excellent regulations don't address the fundamental grid limitations of wind energy identified earlier. Wind energy will not be acceptable until *all* seven criteria are met.

Put another way: wind energy should not be allowed on the public grid until there is scientific proof that it is a net societal benefit.

Does wind energy's abysmal failure mean that all "renewables" are similarly deficient? No. Each alternative power source should be scientifically evaluated. Industrial geothermal holds significant promise. For scientifically based energy information, see WiseEnergy.org.

If we abandon our successful and time-tested criteria for selecting our sources of electrical power and allow lobbyists to dictate our energy policies, there will be incalculable negative impacts on every person on the planet.

References

"Direct Federal Financial Interventions and Subsidies in Energy in Fiscal Year 2010." http://www.eia.gov/analysis/requests/subsidy/. Accessed on May 15, 2014.

"Environmental Impacts of Wind-Energy Projects." http://www.nap.edu/nap-cgi/report.cgi?record_id=11935&type=pdfxsum. Accessed on May 15, 2014.

John Droz Jr. is an independent physicist (energy expert) and longtime (30± years) environmental advocate, who has been extensively involved with energy and environmental matters. He has put together an informal all-volunteer coalition (Alliance for Wise Energy Decisions [AWED]) of some 10,000 individuals, of whom about 700 are PhDs. More information about AWED and John can be found at WiseEnergy.org.

WHY WIND ENERGY IS GOOD FOR AMERICA AND THE WORLD

Tom Gray

Wind energy is a clean, renewable form of energy with enormous economic and environmental benefits for the United States and other countries. Following are some of these benefits:

—*Renewable energy supply:* First, wind energy is "renewable," which means that its supply is essentially inexhaustible. Fossil fuels, as their name indicates, come from the fossilized remains of plants and animals that have accumulated over hundreds of millions of years. When they are burned, a similarly longtime scale is required to replenish them. In 2013, U.S. wind turbines passed a little-noticed milestone—in the 30 years since the first turbines were installed, the wind "fleet" had cumulatively generated as much electricity as could be generated by burning one billion barrels of oil, but the wind resource remained undiminished and ready to produce still more energy.

—*Creation of new domestic manufacturing jobs:* Wind turbines are large, heavy pieces of equipment, roughly comparable to a Boeing 747 jet airliner or bigger. The cost of transporting heavy turbine parts such as blades or towers from where they are manufactured to where they are installed can add as much as 20 percent to the cost of a wind project ("Wind Belt Transportation Improvement Project" 2014, 5), so it makes more business sense to manufacture these parts domestically. This in turn has led to new manufacturing jobs—in 2013, some 560 plants in forty-three states produced one or more components for the wind energy industry.

—*No global warming pollution:* Because wind energy generates electricity through direct conversion of the kinetic energy in the wind, it produces none of the harmful emissions that result from fossil fuel combustion. The most important of these avoidable emissions is carbon dioxide

(CO2), one of the leading "greenhouse gasses" associated with global warming.

—*No air pollution:* Generating electricity with wind energy similarly creates no emissions of sulfur dioxide (SO2), an air pollutant that is a primary component of acid rain; nitrogen oxide (NOx), a primary component of smog; or mercury, a poisonous heavy metal that is emitted from burning coal and which contaminates fish and other animals in the food chain.

—*No water pollution:* The use of hydraulic fracturing ("fracking") to extract natural gas and spills of fossil fuels such as oil results in water pollution. Wind energy uses no fuel except the wind and does not pollute our water supplies.

—*No water use:* Power plants that rely on thermal generation (coal, natural gas, oil, and nuclear) all require large amounts of water for cooling—electricity generation is one of our nation's largest users of water. Hydroelectric generation also involves the loss of substantial quantities of water through evaporation from reservoirs. Wind energy uses virtually no water, a characteristic that will make it exceptionally valuable in a future where global warming is expected to increase the severity of droughts and threaten food supplies.

—*No hazardous (e.g., coal ash) or radioactive waste:* Every year, the nation's coal plants produce 140 million tons of coal ash, a toxic by-product that is dumped in the backyard of power plants across the nation, into open-air pits and surface waste ponds. Nuclear power plants similarly produce radioactive waste that poses a vexing storage problem. Wind uses no fuel and produces no waste.

—No mining or drilling for fuel: Critics of wind energy sometimes point to the large amounts of land required for wind farms, contrasting them with the small footprint of fossil-fired power plants. In doing so, they miss two essential facts: (1) Since wind turbines actually occupy only 2–5% of the land within a wind farm's boundaries, the rest can still be used for farming or ranching, and (2) Large areas of land are disturbed every year by the mining and drilling required to extract fossil fuels.

—No hazards from transportation of fuel: Natural gas pipelines sometimes explode, and trains carrying oil sometimes derail, with catastrophic results. No fuel transportation is needed for wind.

—Rural economic development: Because wind speeds in cities tend to be lower and land is more expensive, wind farms are typically built on open farmland or ranchland in rural areas (or on ridgelines where the terrain is hilly or mountainous). As of the end of 2013, the United States had nearly $120 billion worth of wind turbines installed (AWEA 2014), almost all of them in rural counties which are experiencing difficult economic times (Gray 2014). New wind farms often are the single largest property taxpayers in rural counties, helping boost the tax base and reduce homeowners' taxes. Many counties have built new schools and improved health and public safety facilities with tax revenue provided by wind energy. In addition to tax payments, wind farms provide a modest number of skilled, well-paying jobs (for wind technicians), which are often hard to come by in rural communities.

—Dependable farm income: A farmer or rancher who hosts wind turbines on his or her land can realize land rental payments of $3,000 or more per megawatt (MW) of installed wind capacity (a single turbine is typically 1.5 to 3 MW in size). At the same time, since turbines and their associated equipment and service roads only physically occupy 2–5% of the land, they become an enormously valuable "crop," usually paying more per acre than any other use of farm—or ranchland. The added income helps farmers stay on the land and also provides a diversified income source that is not affected by drought, heavy rainfall, or other weather extremes that can damage crops and reduce yields.

—Stabilization of electricity prices: Because wind energy uses no fuel, it offers electric utilities and their customers an electricity source that is immune to fuel price spikes, such as when unusually cold weather causes high demand for fossil fuels. This can save consumers millions of dollars in fuel costs. In addition, a utility can sign a long-term contract for wind-generated electricity at a guaranteed price, something that is normally not possible for fuel supply. This helps make electricity prices more stable and predictable, benefiting the economy.

References

Gray, Tom. "As Rural Economy Struggles, Wind Power Provides Critical Help." http://aweablog.org/blog/post/As-rural-economy-struggles-wind-power-provides-critical-help. Accessed on May 29, 2014.

"Wind Belt Transportation Improvement Project." [U.S. Department of Transportation].http://www.iowadot.gov/recovery/TIGER/pdfs/UP%20Spine%20Line%20Project/UP%20TIGER%20Wind%20Belt%20Application%209-14-09_final.pdf. Accessed on May 29, 2014.

Tom Gray recently retired after thirty-three years with the American Wind Energy Association, as a consultant or staff member (including fifteen years as Director of Communications). He is a graduate of Haverford (Pennsylvania) College and lives in Norwich, Vermont.

SMALL-SCALE WIND POWER

Yoo Jung Kim

For thousands of years, mankind has harnessed the power of the wind for applications such as propelling ships, grinding grains, and pumping water. In comparison, while generating electricity from wind power is a more recent phenomenon, it has been around since the summer of 1887, when James Blyth, a Scottish academic and electrical engineer, installed a small electricity-generating wind turbine to light his holiday home. This first instance of wind-powered electricity production was also an important step in microgeneration (Redlinger and Andersen 2002).

In microgeneration, individuals, small businesses, and communities generate electricity at a small scale to replace or supplement traditional grid-connected sources. In comparison, commercial wind turbines found in wind farms have a greater individual power output based on their enormous size.

Microgeneration technologies include small-scale wind turbines, which are usually made up of a two- or three-bladed rotor composed of aluminum, fiberglass, or wood, aerodynamically designed to capture kinetic energy from the wind. The rotor is attached to a generator or an alternator mounted on a tower. The amount of wind power captured is proportional to the area of wind swept by the rotor. Based on the purpose of a small-scale wind turbine, a potential consumer has a wide range of choices in the size, the design, and the technical specification (Gipe 2009).

Typically, small wind turbines range from machines producing anywhere from a few watts to 20 kilowatts (kW). While no standardized definition of small-scale wind systems currently exists, one categorization used in Paul Gipe's book *Wind Energy Basics: A Guide to Home-* and *Community-Scale Wind Energy System* distinguishes between micro-turbines, mini-turbines, and household turbines (Gipe 2009).

Across the desert steppes, Mongolian nomads are known to carry micro-turbines on horseback, ranging from a rotor diameter of 0.5 to 1.25 meters. This is usually enough to generate 0.04 to 0.25 kW of electricity, suitable for charging low-power batteries. Mini-turbines are bigger than micro-turbines. They have diameters ranging from 1.25 to 3 meters and can generate from

0.25 to 1.5 kW of power, making mini-turbines suitable for small, off-grid lodgings in remote locations. Finally, some of the biggest small-wind systems are the household turbines. These turbines are sizeable machines, ranging from 3 to 10 meters in rotor diameter and capable of generating power ranging from 1.4 kW to 16 kW (Gipe 2009). According the Department of Energy (DOE), in 2012, nearly 18.4 megawatts of power were being generated through small turbines in the United States alone.

In a year, an average home in the United States consumes about 11,500 kilowatt-hours (kWh) of electricity (about 960 kWh per month) (Installing and Maintaining a Small Wind Electric System 2014). For a demand of this scale, a household would need a mini-turbine or a household turbine to cover a significant portion of the power required. For instance, a wind turbine that can generate 1.5 kW of power would be able to meet the needs of a home requiring 300 kWh per month in a location with an average wind speed of 14 miles per hour (6.26 meters per second) ("Installing and Maintaining a Small Wind Electric System" 2014).

Considering the complexities of wind power, individuals who are interested in setting up a small wind turbine must consider several factors, such as the pattern of their energy expenditure, the daily and seasonal wind flow at the installation site, community and residential zoning requirements, the cost of maintenance, and potential incentives from utilities or the state/federal government. Addressing these questions will allow potential buyers to decide whether installing a small-scale wind turbine will be financially and technically feasible in the long run.

To get a rough estimate of the wind velocity in communities, individuals can use wind resource maps generated by the DOE (Wind Resource Maps and Anemometer Loan Program Data 2014). The highest average wind speeds in the United States are generally found in seacoasts, ridgelines, and in the Great Plains. Individuals can also find information regarding average wind speeds from nearby airports, but because local terrains can significantly influence the direction and the velocity of the wind, the data may not be accurate for a nearby location. Professional small wind site assessors can aid consumers by fine-tuning estimates from wind resource maps and providing the best perspective of the annual electricity output from a small-scale wind system.

If a turbine cannot deliver the total amount of energy required, it can be used in combination with other microgeneration sources—such as photovoltaic solar panels—to create a hybrid power system. Because the most effective operating periods for wind and solar systems take place at different times of the day and the year, hybrid systems can produce additional power more consistently. Furthermore, some microgeneration technology, including small turbines, can be hooked up to the utility-supplied electricity distribution system, also known as the "grid," to provide a stable source of additional power. Under certain situations, if a microgeneration system produces more electricity than is needed, the excess may be sent or sold to a local utility.

Although small-scale wind turbines have existed since the late nineteenth century, they have tremendous potential for future growth. However, because of the relative lack of infrastructure and standardization for small-scale wind systems, individuals must carefully consider a wide range of factors before deciding whether wind power is the right source of alternative energy for them.

References

Gipe, Paul. 2009. *Wind Energy Basics: A Guide to Home—and Community-Scale Wind Energy Systems*, 2nd ed. White River Junction, VT: Chelsea Green Publishing Company.

"Installing and Maintaining a Small Wind Electric System. Energy.gov." http://energy.gov/energysaver/articles/installing-and-maintaining-small-wind-electric-system. Accessed on May 28, 2014.

Redlinger, Robert Y. and Per Dannemand Andersen. 2002. *Wind Energy in the 21st Century: Economics, Policy,*

Technology, and the Changing Electricity Industry. New York: Palgrave.

"Wind Resource Maps and Anemometer Loan Program Data." WINDExchange. http://apps2.eere.energy.gov/wind/windexchange/windmaps/. Accessed on May 28, 2014.

Yoo Jung Kim is a former editor in chief of The Dartmouth Undergraduate Journal of Science and a graduate of Dartmouth College with a major in biology. She is currently writing a how-to guide for college science students for the University of Chicago Press.

WHERE THE WINDS DON'T STOP: HIGH-ALTITUDE WIND ENERGY HAS BIG POTENTIAL

Dave Levitan

The wind turbines that dot the landscape peak at around 300 feet above ground, with the massive blades spinning a bit higher. The wind, however, does not peak at 300 feet. Winds are faster and more consistent the higher one climbs, and pushing wind power up higher into the atmosphere represents an enormous opportunity. There are many challenges, but a growing number of small companies are working hard to get up there within the next few years, with numerous designs and ideas aimed at harvesting wind power high in the sky.

Just how much energy is up there? In two separate analyses, researchers have found an almost unlimited supply of energy at higher altitudes (Jacobson and Archer 2012; Marvel, Kravitz, Caldeira 2013); in other words, we could keep building high-altitude wind turbines until they made up almost all of the world's power supply and we still wouldn't run out.

But those same experts who see all that potential warn that the engineering challenges are huge. How do you safely suspend airborne turbines hundreds or thousands of feet off the ground? How do you keep them aloft for long periods of time in high winds without having to perform frequent, costly maintenance? And what about interference with all the planes that fly around up there?

The basic premise of airborne generation is to tether a device to the ground and let it fly around in the strong winds like a kite, either generating power and sending it down a wire to the ground or using the wire itself to produce electricity on the ground. The specific devices on the end of the tether vary widely in design. Earth-bound windmill design is largely settled, but up in the sky, it seems as if anything goes. There are rigid airplane-like wings outfitted with multiple small turbines; softer kite-like devices that fly in figure eights and generate power by coiling and uncoiling a tether; devices that resemble a blimp rotating around a horizontal axis; and several other concepts. No one is sure which design is the best, though some may be better suited for large wind farms, with lots of devices generating lots of electricity, whereas others may fit better as individual devices in areas that just need a little bit of power.

One company that has made a lot of progress on a high-altitude wind device is Makani Power, based in Alameda, California. Makani's device is known as a "rigid wing"—it resembles an airplane with propellers, though it is connected to the ground by a long wire. It flies in a big circle, and the propellers are turned by the wind to generate electricity, which is sent down the wire to the ground where it can be used. Makani hopes to eventually build big versions of this device; the largest, which is still just an idea, would be as big as a Boeing 747 jet airplane! Something that size would generate enough electricity to power 1,000 homes or more, and it might even end up being cheaper than other power sources.

Ampyx Power, a company in the Netherlands, is also getting close to building wind farms with their devices. The Ampyx idea is to use a glider—again, that resembles an airplane—that pulls on its tether to generate power. Imagine turning a crank to create energy in a generator; the Ampyx glider does that by flying in figure-8 formations as much as 2,000 feet off the ground that pull on the tether. In ideal conditions, these gliders could keep flying for days without any help from humans on the ground.

Another company, called Altaeros, has decided that balloons are the way to go. A helium-filled blimp surrounds what looks like a standard wind turbine; basically, the company uses a balloon to lift a turbine up to 1,000 feet. One of these devices began testing in Alaska in 2014, and earlier tests showed that even at 350 feet off the ground, the turbine held aloft by the balloon generated twice the electricity as that same turbine could make at 100 feet high.

All these and other companies are making good progress, but getting to the point where these are more than just ideas or prototypes will be difficult. One obstacle is how the authorities will regulate the devices. What, exactly, is an 80-foot-wide device, tethered to the ground, flying circles 1,000 feet up in the air? Is it an airplane? A "building" or an "obstacle"? Should the same agency that tells airplanes where to fly also tell the turbines where to fly? No one is yet sure how that will work.

Safety is also a concern. What if that big, 747-sized device Makani wants to build snapped its tether and plunged to the ground? Even with smaller devices, the tether itself is a huge safety risk—thousands of feet of thick metal wire could cause serious damage.

And it's only when we look way up that we see the true potential of high-altitude wind power. All those devices only function in the 500- to 1,000-foot range, about the height of relatively common skyscrapers. But the jet stream is the real prize. Jet streams are fast-moving winds up in the atmosphere, around 5 to 10 miles above the ground. The winds up there never really stop blowing and can reach 100 miles per hour or more; what if we could capture all that energy?

The engineering challenges even at 1,000 feet, though, suggest to most experts that the jet streams are probably out of reach for wind power. After all, if a 1,000-foot cable is dangerous and hard to control, imagine trying to reel in a 40,000-foot cable. But the good news is that there is plenty of energy available even at lower altitudes. If we want to keep pushing for clean energy, pushing the wind turbines up and up and up is a great way to get there.

References

Jacobson, M. Z., and C. L. Archer. 2012. "Saturation Wind Power Potential and Its Implications for Wind Energy." *Proceedings of the National Academies of Sciences*. 9: 15679–15684.

Marvel, Kate, Ben Kravitz, and Ken Caldeira. 2013. "Geophysical Limits to Global Wind Power." *Nature Climate Change*. 3: 118–121.

Dave Levitan is a science journalist. He writes a lot about energy and the environment and thinks that *the sun and the wind won't mind if we use as much of them as we can. Some of this essay first appeared at Yale Environment 360.*

SITING A WIND FARM

John Thomas Post

Finding a suitable location for a wind energy project, from a single turbine to an array of hundreds—a wind farm—is an important step in the evolution of wind energy. Unlike fossil fuels such as coal or gas, which can be shipped or piped long distances, a wind turbine's "fuel," the wind, cannot be transported. Instead, wind turbines must be placed where there is a significant and consistent source of wind.

But siting a wind turbine is not as simple as finding a windy spot. It can be a lengthy, complex process with a variety of ecological, social, economic, and design hurdles. This is one reason why only one in ten initially conceived wind energy projects make it to the construction and operation phases! Some older wind farms, including Altamont Pass in California, have even modified their arrays years after construction, removing or replacing turbines in order to conform to changing environmental standards or community preferences.

Given that wind turbines are engineered to last twenty years or longer, good siting and long-term planning are very important. While every project and location presents its own unique set of challenges to siting, a handful of issues are commonly encountered by developers and must be planned for. The American Wind Energy Association (AWEA) recommends that, when siting a wind turbine, potential developers consider:

- wind resource,
- land or ocean rights,
- federal, state, and local permitting,
- electrical transmission distance,
- and financial considerations, including both investor(s) and purchaser(s).[2]

But depending on the circumstances of a location, there may be other factors to think about. For example, it is increasingly important for developers to consider visual impacts, social acceptance, and cumulative impacts on the environment.[3]

Many organizations publish wind maps that identify areas of high wind resource. Developers use these maps to narrow their search for a suitable location, a process followed by more in-depth surveys with precise weather and wind instruments. A tool called an anemometer is widely used for measuring wind speeds. Anemometers are placed over a prolonged period of time, in order to gauge seasonal variability and total wind resource in a location. Sites with wind speeds greater than 13 miles per hour are generally deemed suitable for commercial-scale wind energy.

Wind is highly variable and can be affected by seasonal changes in weather and climate, topography, and "surface roughness" like trees or tall buildings. For those reasons, the best locations for a wind farm are often in open plains, atop hills, in coastal areas, or offshore, where turbine hubs are significantly above the surrounding landscape.

A wind farm's distance from major roads, population centers (where its electricity is used), and transmission lines (how its electricity is transported) are also important considerations. Developers try to strike a balance between locating a wind farm in an accessible area with a good wind resource, which may be rural, and a more populated area, where the electricity generated can efficiently meet consumer demand.

Once a developer decides on a location with a suitable wind resource, he or she must obtain the appropriate permits from a variety of federal, state, and local agencies. Depending on the specifics of a project, the agencies may include the Federal Aviation Administration (FAA), Army Corps of Engineers, U.S. Fish and Wildlife Service (USFWS), and many others, including state environmental agencies and local zoning boards. For any project receiving federal funding or located on federal lands, the National Environmental Protection Act requires an environmental impact statement or assessment for projects that may have a "significant environmental impact." These permits, consultations, and assessments ensure that a wind farm meets environmental protection standards and does not interfere with human activities.

Wind turbines may have unforeseen ecological impacts, so a wind energy project must be sited to avoid major migratory routes and areas of significant biodiversity or endangered species. Concerns have been raised over bat and bird collisions with rotating turbine blades, and interference with marine life when turbines are sited offshore. However, careful site selection, and research and monitoring of at-risk species can significantly reduce potential problems.

Finally, potential developers must consider a wind farm's impact on human activities. This includes a project's compatibility with nearby land uses and community values and aesthetics. Developers must also minimize risks to public safety and other nuisances such as noise or shadow flicker. These concerns are most often addressed through a public engagement process. Early and frequent involvement of community members, public officials, and economic development and conservation organizations can significantly increase public support for a project and the likelihood of its success.

Siting a wind farm involves much more than finding a windy location. Developers must strike a balance between the wind resource and distance to transmission lines and population centers. They must also take into consideration permitting and a variety of ecological, social, economic, and design challenges. Good siting practices can reduce the potential for long-term negative impacts, promote social acceptance and public involvement, and provide a clean, safe renewable source of electricity for future generations.

In recent years, as wind energy has increased in importance, many of these practices have been standardized and streamlined, allowing developers to focus on emerging challenges such as social acceptance, visual impacts, and cumulative environmental impacts.

John Thomas Post is a researcher in Regional Planning at the University of Massachusetts, Amherst. His interests include offshore wind energy, distributed solar power, and the social and economic networks formed around renewable energies. In his free time, he enjoys bicycling, gardening, and perfecting his hook shot in basketball.

WIND TURBINE SYNDROME: HEALTH EFFECTS OF NOISE FROM LARGE WIND TURBINES

Eric Rosenbloom

It has been known since the early 1980s that noise from large wind turbines can adversely affect human health. In 1981, physicist Neil Kelley and colleagues reported their investigation of complaints from residents living within 3 kilometers of an experimental 2 MW downwind two-blade wind turbine in Boone, North Carolina, which had begun operation in the fall of 1979 (Kelley et al., 1981). Considering that people reported "feeling" the sounds more than hearing them, that the noise was more annoying indoors, that small objects near walls and the glass in picture frames often rattled, and that apparent noise levels were only moderately increased, it seemed to the researchers that infrasound (below the threshold of hearing, or less than 20 Hz) and low-frequency (less than 100 Hz) noise (ILFN) was resonating with the building structures as well as within the subjects' bodies to create feelings of pressure, uneasiness, and vibration. And, indeed, their measurements showed that ILFN pulses dominated the sound energy from the turbine.

Unfortunately for neighbors of large wind turbines, Kelley's research was shelved as commercial wind turbine makers adopted the upwind three-blade design that is usually seen today, asserting that the problem of pulsing (throbbing) low-frequency noise was thereby solved. But commercial models did not approach the size of the one Kelley studied until around 2000, which was, not surprisingly, when many physicians and others, for example, Amanda Harry in England, David Iser in Australia, Robyn Phipps in New Zealand, and Michael Nissenbaum in Maine, started noticing an increase of health complaints after nearby wind turbines began operating. Complaints included headache, dizziness, feeling of pressure, stress, and depression. People experienced relief when they left the area or the turbines weren't operating. Desperation to move away was common, and many families did so if they could. In Ripley, Ontario, the wind energy companies bought several homes of families experiencing health effects (Elma-Mornington Concerned Citizens, 2013).

Nissenbaum compared people who lived within 3,500 feet (just over 1 kilometer) versus people who lived 3 miles away from the twenty-eight wind turbines in Mars Hill, Maine, showing a clear correlation between the wind turbines and health complaints (Nissenbaum et al., 2012). By this time,

French and German health experts, British noise experts, and even a German project developer recognized the need for greater distances between large wind turbines and dwellings than were commonly allowed (Chouard, 2006; Noise Association, 2006; Quambusch and Lauffer, 2008; Retexo-RISP-Marketing, 2004; Villey-Migraine, 2004). The Department for Environment, Food and Rural Affairs in the United Kingdom was aware of the health effects of ILFN and of wind turbines as a source (Leventhall, 2003).

In Portugal, researchers of the effect of ILFN in the body were asked to investigate a home near wind turbines. They found the ILFN levels to be similar to those found at homes near industrial sites, whose residents showed changes to the heart, lungs, and muscles that the researchers called "vibroacoustic disease." Follow-up research of the residence near the wind turbines documented a variety of health effects. Initially, the closest turbine was ordered to be removed and three others to be shut down at night. Eventually those were ordered removed as well to protect the health and well-being of the residents (Alves-Pereira and Castelo Branco, 2007; Supremo Tribunal de Justiça, 2013).

In many places, turbines have been ordered to be shut down at night so that people can sleep. Sleep disturbance itself is considered to be a health effect by the World Health Organization, because sleep is required for both physical and mental health. Combined with the stress from excessive noise, lack of good sleep can lead to long-term problems such as learning disabilities in children, work impairment, and cardiovascular disease (Berglund et al., 1995; Goines and Hagler, 2007).

Meanwhile, physician Nina Pierpont of New York interviewed people from around the world who complained of health effects from wind turbines. She gave the name "wind turbine syndrome" to the common set of symptoms associated with nearby wind turbines: sleep disturbance and deprivation, headache, tinnitus (ringing in ears), ear pressure, dizziness, vertigo (spinning dizziness), nausea, visual blurring, tachycardia (fast heart rate), irritability, problems with concentration and memory, and panic episodes associated with sensations of movement or quivering inside the body (Pierpont, 2009).

Pierpont realized that these symptoms, as well as the fact that people were affected differently—some severely, others not at all—were consistent with inner ear disturbances, akin to motion sickness, that can be caused by noise, especially ILFN (Salt and Hullar 2010). Long-term exposure to high levels of ILFN has long been known to have effects on health, such as those studied in Portugal under the term vibroacoustic disease. Now the likely mechanism for the effects of short-term exposure to wind turbine noise has been found. With the possibility of pulsing ILFN acting on the inner ear to cause the unexpectedly high rate of complaints around wind turbines compared with other sources of noise (Janssen et al. 2011; Pedersen and Waye 2004), acoustic engineers started measuring wind turbine noise in the lower frequencies and rediscovered what Neil Kelley had found almost thirty years before: that noise from large wind turbines is characterized by pulsing ILFN that is associated with complaints and health problems (Channel Islands Acoustics et al. 2012; Møller and Pedersen 2011).

And the solution (at least for human neighbors) is the same: large setback distances to avoid subjecting people to not just increased audible noise as recommended by the World Health Organization (Berglund et al. 1995) but also pulsing ILFN (Kelley 1987; Noise Bulletin 2011).

References

Acórdão do Supremo Tribunal de Justiça (Judgment of the Supreme Court). Revocation of Judgment. 2013. 2209/08.0TBTVD.L1.S1, 7ª Secção, Granja da Fonseca. May 30, 2013. www.dgsi.pt/jstj.nsf/954f0ce6ad9dd8b980256b5f003fa814/4559d6d733d1589780257b7b004d464b. Accessed on April 7, 2014.

Alves-Pereira, Mariana, and Nuno A. A. Castelo Branco. "In-Home Wind Turbine Noise Is Conducive to Vibroacoustic Disease." (abstract). Presented at Wind Turbine Noise Conference 2007, September 20–21, 2007, Lyon, France. http://www.confweb.org/wtn2007/ABSTRACTS_WTN2007.pdf. Accessed on April 7, 2014.

"AM Conditions Approved." Noise Bulletin, Issue 53. http://www.empublishing.co.uk/noise/noise53.pdf. Accessed on April 7, 2014.

Berglund, Birgitta, Thomas Lindvall, and Dietrich H. Schwela, eds. 1995. "Guidelines for Community Noise." Geneva: World Health Organization, 1995. http://whqlibdoc.who.int/hq/1999/a68672.pdf. Accessed on April 7, 2014.

Channel Islands Acoustics, Hessler Associates, Rand Acoustics, and Schomer and Associates. 2012. "A Cooperative Measurement Survey and Analysis of Low Frequency and Infrasound at the Shirley Wind Farm in Brown County, Wisconsin." Wisconsin Public Service Commission. http://docs.wind-watch.org/Shirley-LFN-infrasound.pdf. Accessed on April 7, 2014.

Chouard, Claude-Henri. 2006. "Le Retentissement du Fonctionnement des Éoliennes sur la Santé de L'homme." ("Repercussions of Wind Turbine Operations on Human Health.") lL'Académie nationale de médecine. http://docs.wind-watch.org/FrAcadMed-eoliennes.pdf. Accessed on April 7, 2014.

Elma-Mornington Concerned Citizens. Case Study: Impact of a Wind Turbine Project on a Rural Community. http://docs.wind-watch.org/ripley-case-study.pdf. Accessed on April 7, 2014.

Goines, Lisa, and, Louis Hagler. 2007. "Noise Pollution: A Modern Plague." *Southern Medical Journal.* 100(3): 287–294.

Janssen, Sabine, et al. 2011. "Comparison between Exposure-Response Relationships for Wind Turbine Annoyance and

Annoyance Due to Other Noise Sources." *Journal of the Acoustical Society of America.* 130(6): 3746–3753.

Kelley, N. D. 1987. "Proposed Metric for Assessing the Potential of Community Annoyance from Wind Turbine Low-Frequency Noise Emissions." http://www.nrel.gov/docs/legosti/old/3261.pdf. Accessed on April 7, 2014.

Kelley, N. D., H. E. McKenna, and R. R. Hemphill. 1981. "A Methodology for Assessment of Wind Turbine Noise Generation." *Journal of Solar Energy Engineering.* 21: 341–356.

Leventhall, Geoff. 2003. *A Review of Published Research on Low Frequency Noise and Its Effects.* London: Department for Environment, Food and Rural Affairs, May 2003.

Møller, Henrik, and Christian Sejer Pedersen. 2011. "Low-Frequency Noise from Large Wind Turbines." *Journal of the Acoustical Society of America.* 129(6): 3727–3744.

Nissenbaum, Michael A., Jeffrey J. Aramini, and Christopher D. Hanning. 2012. "Effects of Industrial Wind Turbine Noise on Sleep and Health." *Noise & Health.* 14(60): 237–243.

Noise Association. 2006. "Location, Location, Location: An Investigation into Wind Farms and Noise." http://www.countryguardian.net/Location.pdf. Accessed on April 7, 2014.

Pedersen, Eja, and Kerstin Persson Waye. 2004. "Perception and Annoyance due to Wind Turbine Noise—A Dose-Response Relationship." *Journal of the Acoustical Society of America.* 116(6): 3460–3470.

Pierpont, Nina. 2009. *Wind Turbine Syndrome.* Santa Fe, NM: K-Selected Books, 2009.

Quambusch, E., and M. Lauffer. 2008. "Infraschall von Windkraftanlagen als Gesundheitsgefahr." ("Infrasound from Wind Turbines as a Health Hazard.") ZFSH/SGB-Zeitschrift für die sozialrechtliche Praxis, August

2008. http://docs.wind-watch.org/infraschall-Quambush-Lauffer-200808.pdf. Accessed on April 7, 2014.

Retexo-RISP-Marketing. "Importan [*sic*] Factors when Planning a Wind Farm." www.retexo.de/english/wind/seite5a.htm. Accessed on April 7, 2014.

Salt, Alec N., and Timothy E. Hullar. 2010. "Responses of the Ear to Low Frequency Sounds, Infrasound and Wind Turbines." *Hearing Research.* 268(1–2): 12–21.

Villey-Migraine Marjolaine. "Eoliennes, Sons et Infrasons: Effets de L'éolien Industriel sur la Sante des Hommes." ("Wind Turbines, Noise, and Infrasound: Effects of Industrial Wind Turbines on Human Health.") Université Paris II-Panthéon-Assas. http://docs.wind-watch.org/villey-migraine_eoliennesinfrasons.pdf. Accessed on April 7, 2014.

Eric Rosenbloom is a medical sciences editor and writer. He has been president of National Wind Watch, a clearinghouse for news and research concerning industrial-scale wind power, since 2006.

THE CAREFUL PLANTING OF OFFSHORE WIND FARMS

Linda Zajac

It's imperative that the United States take strong and swift action on climate change. Left unchecked, the harmful effects of rising carbon dioxide levels will only keep increasing. In the United States, offshore wind power is a vast and untapped source of renewable energy. Offshore wind farms can reduce our reliance on fossil fuels, cut our carbon emissions, decrease our dependence on the Middle East, and create jobs.

Out at sea, the wind is stronger and blows more consistently than it does over land. An endless supply of wind can power turbines for years. Although there is a pressing need for renewable energy, offshore wind farm installation should not be hasty and haphazard. These projects require careful construction to ensure that wildlife isn't harmed.

The European Union is a world leader in wind energy. Many member countries are aggressively seeking to meet their carbon reduction targets. In the United Kingdom, neat rows of giant pinwheels tower high above the sea. These slender turbine blades whirl in the wind, generating clean electricity. The United Kingdom alone has over one thousand offshore wind turbines. Denmark gets nearly one-third of its electricity from the wind. At the end of 2013, eleven European countries were operating over 2,000 offshore turbines (Corbetta 2014). In contrast, the United States had no offshore wind turbines in operation.

The biggest hazard to marine mammals arises during installation of these structures. To anchor offshore turbines, energy companies hammer steel piles into the seafloor. The extremely loud noises from pile driving can be harmful to marine animals that use sound to communicate, navigate, and find food. The noise may damage their hearing, alter their movements, or cause them to avoid preferred feeding grounds.

In 2006, when Talisman Energy installed two wind turbines off the coast of Scotland, researchers measured the impact of pile-driving noise on marine mammals. The site was about 16 miles from a marine-protected area for bottlenose dolphins. Harbor porpoises, minke whales, and common and grey seals also frolicked in the waters. Dr. Helen Bailey of the Chesapeake Biological Laboratory, University of Maryland, and Dr. Paul Thompson of the University of Aberdeen, the United Kingdom, conducted several studies. Before construction got under way, the team used hydrophones to measure the existing noise in the ocean (Bailey et al. 2010).

To lessen the impact on marine animals, Talisman Energy hired trained mammal observers and proceeded slowly with pile driving. Observers scanned the sea with binoculars to ensure there were no marine mammals within 1 kilometer of the drilling area. To erect each turbine, workers lowered a steel plate onto the seafloor. Four steel piles secured the corners of each metal plate. To alert animals and allow them time to flee, workers started with five strokes of the hammer at the lowest setting. Gradually, after twenty minutes, they pounded at full force. They hammered the hollow piles until they were nearly flush with the seafloor, a process that took about two hours per pile (Bailey et al. 2010).

While the piles were anchored in place, the scientists again measured the noise with hydrophones. They took readings at various distances from the installation site. Bailey and Thompson learned that noise from pile driving could be detected up to 44 miles away. At 31 miles from the turbine, the sound was loud enough to alter the behavior of bottlenose dolphins and minke whales (Bailey et al. 2010).

Harbor porpoises are highly sensitive to noise. To determine if they were present during pile driving, the scientists deployed two Timed Porpoise Detectors, or T-PODS. These devices recorded echolocation clicks before, during, and after pile installation. The echolocation clicks of porpoises dramatically declined while the piles were being hammered (Thompson et al. 2010).

The study in Scotland shows that pile driving can be heard over long distances, and it does affect marine mammals. The potential for harm should not be a reason to abandon a worthy wind project. By doing research studies to assess the impact on wildlife and by utilizing care and creative problem solving during installation, these harmful effects can be minimized.

Climate change is a global problem, and it will take a global effort to reduce carbon dioxide levels. It's shameful that we haggle here in the United States over whether climate change is real or manmade while other countries are making huge strides reducing their carbon emissions. Efforts to misrepresent the facts, divide the issue with politics, and discredit scientists have worked in favor of the fossil fuel industry. In a review of over 11,000 climate research papers, over —97 percent of those that took a position on the cause of climate change point to manmade causes (Cook et al. 2013). There is no debate. Americans need to mend their gas-guzzling, carbon-spewing, and resource-wasting ways.

An investment in offshore wind farms is an investment in the health of our nation. The value is priceless. To tackle the greatest challenge of the century, the United States needs to plant a crop of offshore turbines, making sure their roots don't harm the wildlife that fill the waters.

References

Bailey, Helen, et al. 2010. "Assessing Underwater Noise Levels during Pile-Driving at an Offshore Windfarm and Its Potential Effects on Marine Mammals." *Marine Pollution Bulletin*. 60: 888–897. http://abdn.ac.uk/lighthouse/documents/Bailey_Assessing_underwater_2010_MPB.pdf. Accessed on May 13, 2014.

Cook, John, et al. 2013. "Quantifying the Consensus on Anthropogenic Global Warming in the Scientific Literature." *Environmental Research Letters*. 8(2): 024024.

Corbetta, Giorgio et al. 2014. "The European Offshore Wind Industry—Key Trends and Statistics 2013." January 2014. http://www.ewea.org/fileadmin/files/library/publications/statistics/European_offshore_statistics_2013.pdf. Accessed on May 13, 2014.

Thompson, Paul M., et al. 2010. "Assessing the Responses of Coastal Cetaceans to the Construction of Offshore Wind Turbines." *Marine Pollution Bulletin*. 60(8): 1200–1208.

Linda Zajac is an award-winning freelance science writer. She pursues captivating stories about scientists who use cutting-edge technology to advance medicine, study wildlife, and protect the environment. Zajac is a past student and teacher's assistant for Climate Literacy, a course offered by the University of British Columbia.

Notes

1. In contrast, in the non–English speaking countries, notably Denmark and Germany, opponents of wind energy failed

to take up the issue. For example, an article in Der Spiegel, the top-selling Germany weekly, "Green Headache—Resistance Mounts to Germany's Ambitious Renewable Energy Plans"—discussed several reasons why Germans were opposing wind turbines. Health was not even mentioned. Still today, German and Danish Web sites and mainstream media pay little attention to the health issue, despite the fact that Nina Pierpont's book was translated into German soon after publication.

2. Purchasers are often a utility company, or, for larger projects, multiple utilities, which sign a power purchase agreement (PPA), a contract that prices the cost of energy at a certain rate per kilowatt-hour (kWh).

3. In the United States, wind energy projects have faced barriers from lack of social acceptance or political support. Even some European nations with widespread deployment of wind energy have experienced backlash from its extensive use. Early and frequent public involvement in the siting process may help ensure timely development of a project and prevent lengthy lawsuits. As wind energy grows in importance, cumulative environmental impacts, which are alterations to the natural environment beyond the realm of an individual project, are also of concern.

DOE/NREL

4 Profiles

Introduction

The history of the growth and development of wind power can be told, at least in part, by the stories of individuals and organizations who have contributed to that process. A complete list of such biographies is well beyond the constraints of this book. This chapter, however, provides a taste of the work and accomplishments of a sample of those individuals and organizations.

American Wind Energy Association

1501 M St., NW, Suite 1000
Washington, DC 20005
Phone: (202) 383–2500
Fax: (202) 383–2505
Email: None provided
URL: http://www.awea.org/

The American Wind Energy Association (AWEA) was founded in Detroit, Michigan, in 1974 by Allan O'Shea, who was a salesman for solar and wind equipment. The association held its first national conference a year later at the University of Colorado in Boulder,

---

Two Westinghouse wind turbines, originally used on a farm in Oahu, Hawaii, were moved to the National Wind Technology Center and were installed on a Mesa near Rocky Flats, where they have been heavily modified for research. The 120-foot-tall turbines are each capable of generating 600 KW of electricity in a 30 mph wind. (Andy Cross/The Denver Post/Getty Images)

with about 125 attendees. The goal of the organization is to bring together leaders in the young wind energy business from around the world. In 1978, AWEA moved its headquarters from Detroit to Washington, D.C., where it could be closer to legislative leaders, who were the target of its lobbying and educational programs. The organization continues to hold annual meetings under the title of WINDPOWER (with the relevant year following, such as WINDPOWER 2014).

AWEA has a somewhat complex system of membership, consisting of corporate, exhibitor, utility, academic, and non-profit levels, with further divisions within each level. Corporate level, for example, is subdivided according to sales income into basic membership, industry participants, established industry participants, fully integrated players, key industry players, industry leaders, and strategic industry leaders. Some of the more than 1,200 organizations that hold membership in AWEA are American Clean Skies Foundation; Axis Crane; Barco Rent A Truck; Beijing Victory; Electric Co., Ltd.; Blade Dynamics; Capital City Renewables LLC; Clinton Community College; Customized Energy Solutions, Ltd.; Doble Engineering Company; Ensemble Wind; Extol Wind; Fabrication Delta; Freudenberg Sealing Technologies; Graham County Economic Development; Guangzhou HY Energy Technology Co, LTD; Herzing University; IWR México; Marten Law; Michigan Technological University; National Wildlife Federation; No Fossil Fuel; Northern Arizona University; Ocean Offshore Energy; RBB Engineering; Saint Lawrence Seaway Development Corporation; Sustainable Energy Strategies, Inc.; Union Of Concerned Scientists; United Brotherhood of Carpenters and Joiners of America; University of Minnesota Duluth; Windrush LLC; Yale University Library; and Zhuji Hiest Motor Co., Ltd. The organization also has regional partners in every part of the nation except the Southwest, including the Alliance for Clean Energy New York, CalWEA (California), Center for Energy Efficiency and Renewable Technologies (California), Interwest Energy Alliance (Southwest), Mid Atlantic Renewable Energy

Coalition, Renewable Northwest Project, the Wind Coalition (lower Midwest), and Wind on the Wires (upper Midwest).

AWEA divides its projects into five major categories: educating legislators and the general public about wind energy issues; advocating for legislative and policy actions with the U.S. Congress and at the state level; hosting the annual WINDPOWER conference; sponsoring and conducting research on wind power; and providing opportunities for interested individuals and organizations to become involved in the campaign to promote and expand wind power. At the present time, the association is focusing on three major areas of interest: small wind systems, offshore wind facilities, and community wind programs. It has also identified six major areas in which it carries on its research, advocacy, and educational programs: utilities, operation and maintenance, siting, safety, transmission, and sectors. Each of these areas is further divided into a number of topics of interest. For example, siting programs may focus on topics such as wildlife, public lands, property values, sound and public health, and radar and airspace, all subjects of interest to the wind industry and potential or actual sources of conflict about the development and use of wind programs.

In addition to the annual WINDPOWER conference, AWEA sponsors a second annual conference focusing specifically on offshore wind facilities called OFFSHORE WINDPOWER. It also holds a number of seminars and Webinars on a variety of topics, such as finance and investment, resource assessment, legislative issues, research findings, business strategies, and market trends. AWEA also provides a number of essential publications on the topic of wind energy, most important of which may be the annual *Wind Industry Market Report.* Some examples of other fact sheets, market reports, articles, position statements, testimonies, brochures, position papers, newsletters, regulatory comments, and other publications are Wind Industry Basics (fact sheet), "Offshore Wind: Major Milestones & Achievements" (article), AWEA Comments on USFWS Eagle Guidance (regulatory comment), *The Hidden*

Costs of Energy: National Academy of Sciences Study (Report), What Does Wind Power Mean for America? (brochure), and *Manufacturing Supply Chain Handbook* (handbook). The associate also sponsors a blog on wind issues, "Into the Wind," which provides a speakers' bureau for interested groups, offers a communicators toolkit for those individuals who wish to make presentations about wind energy, and publishes press releases on a number of topics and issues of relevance to wind power development.

Albert Betz (1885–1968)

In 1919, Betz published a historic paper, "Das Maximum der Theoretisch Möglichen Ausnutzung des Windes durch Windmotoren" ("Theoretical Limit for Best Utilization of Wind by Wind Motors"), in which he revealed calculations showing that the maximum amount of energy that can be extracted from the wind by a wind turbine is 16/27 (59.3 percent) of the kinetic energy of the wind. This result is independent of the type of wind turbine involved and is important in wind technology because it gives the maximum amount of energy of any wind turbine for a given wind speed. This value is now called the Betz criterion, the Betz equation, the Betz limit, or Betz's law. In actual practice, modern wind turbines commonly generate no more than about 85 percent of the Betz maximum, and lower values are very common.

Albert Betz was born in Schweinfurt, Germany, on December 25, 1885. He studied philosophy at the Catholic University of Eichstätt-Ingolstadt before attending the Technical University of Munich. He then left school for two years from 1905 to 1906 to work in a shipyard in Kiel before returning to complete his degree in naval engineering at the Technische Hochschule Berlin-Charlottenburg, which he received in 1910. Betz then accepted an appointment as a research assistant at the Institute of Applied Mechanics at the University of Göttingen. During this time he continued his studies in physics and mathematics and qualified for his doctor of philosophy degree from Göttingen in 1919. He then became a member of the faculty at Göttingen and in 1935 was appointed full professor of physics.

While at Göttingen, Betz also served as a scientific member of the Kaiser Wilhelm Institute for Fluid Dynamics, also acting as director of the institute from 1938 to 1947. In 1937, Betz was also appointed to succeed the renowned Ludwig Prandtl as head of the Aerodynamics Laboratory at Göttingen. He continued in that post until 1947, when the laboratory was moved from the Kaiser Wilhelm Institute to the newly created Max Planck Institute for Fluid Dynamics. In addition to his theoretical work at Göttingen, Betz conducted studies on the properties of wind turbine components to determine the structures best suited for energy production. His work largely came to a halt, however, as the onset of World War II transformed priorities at every level of the German government and the success of fossil fuels made studies of wind energy seem somewhat frivolous.

It was not until the 1970s that interest in wind power again reappeared. By that time, however, Betz was no longer on the scene, having died in Göttingen on April 16, 1968. During his life, he received a number of honors and awards, including the Carl Friedrich Gauss Medal of the International Mathematical Union and the German Mathematical Society in 1965.

James Blyth (1839–1906)

Blyth is credited with having built the first windmill in the history of the generation of electricity, hence the world's first wind turbine. While a professor at Anderson's College in Glasgow (now Glasgow and West of Scotland Technical College), Blyth built his windmill to provide electrical power for his vacation cottage in Marykirk. The device consisted of a central post 33 feet tall, to which were attached four arms, each containing canvas sheets 13 feet long. The whole structure was supported by a tripod frame consisting of three wooden posts. Electricity generated by the windmill was used to charge batteries, which, in turn, supplied electrical current for lighting the cottage. A moderate wind was sufficient to light ten 25-volt light bulbs and other small appliances in the cottage. His success with the Marykirk machine prompted Blyth to apply for a patent for his invention, which he received in 1891, and to build a second device to provide electrical power to the local Royal Lunatic Asylum, Infirmary, and Dispensary of Montrose, a machine that continued operating for three decades. Blyth's original Marykirk windmill was dismantled in 1914.

Although a successful demonstration of the potential of wind power as a source of electrical energy, Blyth's machine never became commercially popular. Investors thought that wind power could never be economically competitive with electricity provided by other sources, such as fossil fuel generation, and, in fact, the next wind turbine was not constructed in the United Kingdom until more than sixty years later, in 1951.

James Blyth was born on April 4, 1839, in Marykirk, a small village in Kincardineshire, near the town of Montrose. He attended the Marykirk village school and Montrose Academy before matriculating at the General Assembly Normal School in Edinburgh in 1856. He left school in 1858 to take a teaching position at Claverton Lodge in Bath, England, but returned to Scotland to complete his education at the Edinburgh University in 1859. He received his bachelor of arts degree from Edinburgh in 1861 and accepted a teaching position at Morrison's Academy, Crieff, Perthshire. He left Morrison's in 1870 to participate in the founding of George Watson's College in Edinburgh, where he remained for the next decade. In 1880, he accepted an appointment as professor of natural philosophy at Anderson's College, a post he held for the rest of his life.

At Anderson's College, Blyth came into contact with Sir William Thomson (Lord Kelvin), who, like Blyth, was very much interested in the use of wind power to generate electrical current. The two apparently shared ideas about the construction of a device for achieving this objective, which saw the light of day as Blyth's windmill. Thomson was probably one of the relatively few scholars who recognized the potential of wind power for electrical generation. At one point, he observed, "Even now, it is not utterly chimerical to think of wind superseding coal in some places for a very important part of its present duty—that of giving light."

As successful as it was, Blyth's windmill was not without its problems, the most serious of which was probably its lack of a shutdown mechanism in strong winds. He apparently solved this problem to his satisfaction with his second machine, built in Montrose. But, in any case, circumstances were such that wind power could in no way compete with the inexpensive generation of electricity by coal at the time.

In addition to his interest in wind power, Blyth pursued his studies involving telephone communications, the microphone, and other electrical devices. He died of apoplexy at his home in Glasgow on May 15, 1906. Following his death, his students endowed the Blyth Memorial Prizes in his honor and erected a plaque at Anderson's College in his honor.

Charles Francis Brush (1849–1929)

Brush is best known in the wind industry for his invention of a large wind turbine in 1888. He built the turbine on the grounds of his home in Cleveland, Ohio, where it provided electrical power to the home and surrounding buildings. The turbine consisted of a 60-foot tower made of wrought iron, to which was attached a rotor containing 144 blades with a total area of 1,800 square feet. The wheel rotated at a maximum speed of 500 revolutions per minute. Rotation of the wheel produced an electric current that was used to charge 408 batteries in the basement of the Brush home. The turbine itself remained in operation until 1909, and the batteries were still in use as late as 1929.

Charles Francis Brush was born in Euclid Township, just east of Cleveland, on March 17, 1849. He grew up on his family's farm in Euclid and developed an interest in electrical phenomena as a young boy. He is said to have built electrical devices at his home even before reaching high school age and continued his research on arc lamps while a student at Cleveland's Central High School. Arc lamps had been invented in 1801 by English physicist and chemist Sir Humphry Davy, although it was some years before they reached a stage at which they could be used as a practical source of light. After graduating from high school, Brush enrolled at the University of Michigan, where he majored in mining engineering. He received his degree in 1869, having completed the typical four-year course of study in two years.

Brush then returned to Cleveland, where he began work as an analytical and consulting chemist. He was not very successful in this endeavor and decided to join forces with a boyhood friend, Charles Bigham, to establish a business marketing pig iron and iron ore. This business turned out to be very successful, allowing Brush to spend more of his free time on his true passion—research on electrical devices. He began work on improvements in the design and operation of the electric arc lamp and received a number of patents for his designs. By 1877 he had decided to give up the iron marketing business and devote all his energies to research. In the same year, he received his first patent, "Improvement in Magneto-Electric Machines," for an improved design of an electrical dynamo. He then turned his full attention to the development of arc lamp systems.

Brush's improvements in the arc lamp included modifications of the electrodes for longer life, regulators for the control of lamp operation, and an automatic shut-off system for the lamps. The first successful application of Brush's arc lamp design occurred in 1878, when one of his lamps was installed at a doctor's home in Cincinnati. It took a remarkably short time before individuals, companies, and municipalities to realize the value of this new type of lighting. Brush was commissioned by the cities of Baltimore, Boston, Buffalo, Montreal, New York, Philadelphia, and San Francisco, to name but a few, to hire Brush to design and install arc lamp systems for their communities.

The success of these ventures provided Brush with the income needed to spend his time on whatever activity in which he was most interested, which included wind turbines. It was during this period that he began to study the design of wind turbines and to begin planning for the installation of the large wind turbine at his own home. Unlike his other ventures, however, the wind turbine success was not repeated at other locations in Cleveland or in other cities across the country.

Throughout his life, Brush remained interested in a wide variety of scientific topics, not always with the greatest success. In 1898, for example, he reported that he had found a new gas in the atmosphere which he called *etherion.* The gas appeared to be largely unreactive, and some later biographers have equated it with helium, claiming that Brush had discovered helium in Earth's atmosphere. As it happens, Brush was wrong, and the gas he claimed to have discovered was actually, and most likely to have been, water vapor.

Brush also became interested in the study of gases, both from a theoretical and practical standpoint, and was a financial backer of the Linde Air Products Company. He and his son also formed a company in 1921, Brush Laboratories Company, for the purpose of making and selling beryllium metal and Rochelle salts.

As he grew more successful financially, Brush became an active philanthropist for a number of Cleveland institutions, including Case School of Applied Science, Western Reserve University, University School, the Cleveland School of Art, Lakeview Cemetery, the Cleveland Museum of Art, Trinity Cathedral, and the Cleveland Orchestra. In 1928, Brush established the Brush Foundation in memory of his son, Charles F. Brush, Jr. One of his major gifts was given for the purpose of funding research in the field of eugenics and to study the problems of human overpopulation.

Brush was honored with most of the major awards given for achievements in electrical research, including the Rumford

Prize of the American Academy of Arts and Sciences (1899), the Edison Medal of the American Institute of Electrical Engineers (1913), and the Franklin Medal of the Franklin Institute (1928). He also received honorary degrees from Western Reserve University (1880), Kenyon College (1903), the University of Michigan (1912), and the Case School of Applied Science (1928). In 1881 he was given the French Legion of Honor.

Brush died at his home in Cleveland on June 15, 1929.

William Cubitt (1785–1861)

Cubitt was a talented inventor who developed a form of sails for windmills, later given the name *patent sails*. These sails were made of wooden slats rather than canvas, as had been the practice previously. The sails were fitted with a device that allowed them to change their orientation toward the wind automatically, thus saving considerable time for the mill operator. Prior to their availability, an operator would have to stop the mill every time the wind changed direction, altering the canvas sails to capture the maximum amount of wind possible. Cubitt's patent sails were actually patented in 1813 but never became very popular until the patent expired in 1821. They then became the standard for windmills essentially worldwide until the appearance of the modern wind turbine.

William Cubitt was born in Bacton Wood, near Dilham, Norfolk, England, in 1785. His father was a miller, and young William showed a talent for working in the mill at an early age. He obtained the minimal primary education available to him in his home village and later moved with his family to the village of Southrepps. At the age of fifteen, Cubitt was apprenticed to one James Lyon, a cabinet maker at Stalham in a relationship that was not very satisfactory. Cubitt left the apprenticeship after four years and returned to work with his father at Bacton Wood Mills. Here he began to show his interest and skill at invention, designing a machine for splitting hides. While at

Bacton Wood, Cubitt produced the invention for which he is perhaps best known today, the patent windmill. At about the same time, he opened his own business as a millwright in the village of Horning.

His business at Horning was not very successful, and in 1812 he joined the firm of Ransome & Son in Ipswich, ironfounders, as an engineer. Ransome's major business was the manufacture of plows and other agricultural machinery. During his time at Ransome, Cubitt proved to be an effective manager, expanding the company's business into a wider field of iron manufacturing. For example, the company was hired to build a number of iron bridges, the most substantial of which was the Stoke Bridge in Ipswich. That bridge was opened in 1818 and lasted until 1924. Cubitt remained with the Ransome company as a full-time employee until 1814 and as a part-time employee for four more years.

After Cubitt left the Ransome company, he was involved in a variety of commercial operations. In 1817, for example, he was hired by the city of Ipswich to design and supervise the construction of a system for lighting residents' homes and businesses with gas. He also moved on to supervise and consult on projects involving the construction of new canals and railroad systems in the rapidly industrializing central England.

One of the inventions for which he is perhaps most famous, after his work on windmills, was that of the prison treadmill. On a visit to the prison at Bury St. Edwards, the warden mentioned that he wished that some method could be devised to keep his prisoners better occupied. At the time, prisons had little or no required physical activity and spent most of their time lying around unproductively. Looking for a solution to this problem, Cubitt invented a treadmill that could be operated by groups of prisoners working together. The treadmill was attached to water pumps, mills, or other mechanical devices that could be used for productive labor. The treadmill soon became popular among wardens, and it was adopted by most of the prisons across England. As it happens, some wardens saw another use for the treadmills as a form of punishment or restriction. Prisoners were forced to ride the treadmill, which was not, in fact, attached to anything at all. Instead of being a tool for productive work, they became one more element of punishment.

In another of his signature achievements, Cubitt was involved in the design of the Great Exhibition building in Hyde Park in 1851. His work was so successful on the project that he was knighted by Queen Victoria. He had also been elected a fellow of the Royal Society in 1830 and was a member of a number of other scientific organizations. He retired from active participation in business in 1858 and died at his home in Clapham Common, Surrey, on October 13, 1861.

Georges Jean Marie Darrieus (1888–1979)

Darrieus is best known today for his invention in the early 1930s of a vertical-axis wind turbine. Although vertical-axis windmills had been known and used for many centuries, Darrieus's invention was one of the first such devices to be described in the twentieth century. Darrieus obtained a U.S. patent for his invention on December 8, 1931 (patent number 1,835,018), five years after he had filed his application in Washington. He had filed a similar patent application in Paris a year earlier in 1925. In his patent application, Darrieus described a number of problems associated with the traditional horizontal-axis wind turbine and explained how his design avoided some of those problems:

> The present invention has for its object a turbine in which the blades which receive the action of the wind always present the same surface to the direction of the relative wind. . . .
>
> It is thus possible to give these blades a stream line section analogous to that of the wings of birds, that is to say, offering the minimum resistance to forward movement and capable of converting into mechanical energy

> the maximum available amount of energy of this fluid. . . . (From the patent application; http://s3.ptodirect.com/1835018.pdf)

Although they have traditionally been much less popular than horizontal-axis wind turbines, Darrieus's machines have, nonetheless, found some applications in distributed wind systems and, in some cases, larger commercial systems.

Georges Jean Marie Darrieus was born in Toulon, France, on September 24, 1888. He attended the Central École des Arts et Manufactures in Paris, from which he received his engineering degree in 1910. He then continued his studies in physics at the Electrotechnical Institute of Toulouse. In 1912, he accepted a position at the electrical engineering firm of Compagnie Électro-Mécanique (CEM), a subsidiary of Brown, Boveri & Cie. Darrieus remained with CEM for the rest of his professional career, eventually reaching the position of scientific director of the firm. Most of his research and published papers focus on technical problems associated with the design of wind turbines to obtain maximum conversion of wind energy to electricity.

Darrieus was elected a member of the French Academy of Sciences in 1946 and was awarded the degree of Doctor Honoris Causa from the Polytechnic School of Zurich in 1964. He died on July 15, 1979, in Lyon, France.

Distributed Wind Energy Association

P.O. Box 1861
Flagstaff, AZ 86002
Phone: (928) 255-0214
Email: info@distributedwind.org
URL: http://distributedwind.org/

The Distributed Wind Energy Association (DWEA) was founded in 2010 as an offshoot of the American Wind Energy

Association (AWEA) for the purpose of increasing attention to the special problems, issues, and opportunities associated with distributed wind energy systems. The term *distributed* in this context refers to small wind energy systems, usually with wind turbines of less than 100 kW capacity. The primary objectives of DWEA are to work on regulatory issues involving small wind systems and to better educate policymakers, energy specialists, and general citizens about the potential that small wind systems hold for the nation and the world. The young organization has a relatively small staff overseen by a board of directors of ten wind company executives and employees, led by president Mike Bergey, a mechanical engineer and co-founder of BWC, the Bergey Wind Company. Bergey is also twice past president of the American Wind Energy Association and a longtime activist in the promotion of wind energy in the United States.

Examples of some of the events that DWEA has sponsored or been involved in are the DWEA Business Conference held in February 2014 in Washington, D.C.; the BCSE (Business Council for Sustainable Energy) 2014 Factbook Congressional Briefing, also held in February 2014 in Washington; the Fifth World Summit for Small Wind, held in Husum, Germany, in March 2014; and the Alaska Wind Integration Workshop, held in Fairbanks in March 2014.

At the moment, the most useful service provided by DWEA might be its excellent reference page on federal and state policies on small wind generation and on zoning and permitting procedures for small wind projects. The Web page on this topic has links to or downloads of virtually all essential documents on these topics, such as proposed legislation to extend investment tax credits to small wind facilities, proposed U.S. Fish and Wildlife Service regulations on small wind systems, and distributed wind programs under way in a number of states and provinces such as California, Hawaii, Iowa, New York, and Ontario. The organization's Web site also has a useful section on news relating to distributed wind energy issues in the United States and Canada.

Anton Flettner (1885–1961)

Flettner was a prolific and imaginative inventor who developed a number of new models for flying and sailing ships. He is best remembered in the field of wind power for his attempts to harness the energy of wind to propel ships. To do so, he developed a device that has become known as the Flettner rotor, a large cylinder fixed so as to rotate on its own axis. In his first model, the two cylinders he used were 3 meters in diameter and 15 meters tall, affixed fore and aft of midships on a converted schooner called the *Buckau.* When the wind blew, the cylinders turned at a speed of 125 revolutions per minute, driven by two 11-kilowatt electric motors powered in turn by a 34-kilowatt generator.

Flettner's invention was based on a scientific principle that had been known for many decades known as the Magnus force. Flettner calculated that the force produced by capturing the wind with rotating cylinders was about ten times the force captured by a set of standard sails that would normally have been used on the ship. By 1925, the *Buckau* had been put into service on sixty-two trial voyages hauling coal and lumber around the Baltic Sea and as far west as Scotland. In August 1926, the ship, now named the *Baden Baden,* sailed across the Atlantic Ocean and into New York harbor, where it was welcomed, according to an article in the August 1926 issue of *Popular Science Monthly,* by crowds of onlookers who saw the ship as "a mysterious 'freak' of invention." The magazine went on to say that the ship design was "still in its babyhood . . . [but] [i]ts possibilities seem almost limitless."

Enthusiasm for Flettner's invention gradually faded as engineers determined that the amount of energy obtained from the wind was less than the amount of energy needed to operate the cylinders. The *Baden Baden* was eventually converted back to a conventional sailing ship, and it sank in the Caribbean Sea in 1931.

Anton Flettner was born in Eddersheim (now a district of Hattersheim am Main), Germany, on November 1, 1885. His father was Peter Flettner, a very successful ship builder and owner who had built one of the most prominent homes in the Eddersheim region. After graduation, he taught briefly at Lorsbach and Niederrad, but his primary interest was in the field of invention. During World War I, he worked on the development of a steerable torpedo, a weapon that was at the time unavailable. When he presented his idea to the German Imperial Navy Office, it was rejected because authorities there thought the concept was unworkable.

At the end of the war, Flettner decided that he was no longer interested in teaching and devoted all of his time to invention. He was able to obtain an appointment at the Aerodynamic Research Institute at Göttingen, where he worked with the famous Ludwig Prandtl. His major project between 1922 and 1926 was the wind-driven sailing ship described earlier. He later described the results of this effort in his 1926 book *My Way to the Rotor* (Leipzig: Hase & Koehler-Verlag).

During the later 1920s, Flettner was also working on a somewhat modest invention, which, ironically, is responsible for any fame and name recognition he may retain today: the Flettner rotary ventilator. This simple device uses wind power to activate a horizontal fan that circulates air and other gases in and out of closed spaces. It is widely used today for closed vehicles (most commonly seen on the top of buses) and factories (where they are used to remove noxious and offensive gases).

At the same time, Flettner was moving into another, much more ambitious, area of research and invention: aerodynamics and the design of airships. Partially in response to requests by the German military, Flettner began work in the early 1930s on helicopter designs. He eventually produced a model that contained a pair of horizontal propellers set slightly off-parallel to each other that was very effective in providing the aircraft with stability in flight. His company, Anton Flettner, Flugzeugbau GmbH, produced 24 "Hummingbird" helicopters of this design, but his plans to build more than 1,000 of the aircraft ended when his factory was destroyed by Allied bombing.

At the conclusion of World War II, Flettner was removed to the United States in a program known as Operation Paperclip, in which many of Germany's best scientists and engineers were put to work on American projects. He founded the Flettner Aircraft Corporation, with plans to produce Hummingbird-like helicopters, a project that ended in failure. He continued to contribute to the development of the American helicopter industry, however, applied for and received a number of patents in the field, and eventually contributed to the development of the popular Kaman helicopter. Flettner died in New York City on December 29, 1961.

GE

3135 Easton Turnpike
Fairfield, CT 06828
Phone: (203) 373–2211
Email (for wind services): http://www.ge-energy.com/contact.jsp
URL: http://www.ge-energy.com/wind

According to a 2012 report by Navigant Research, General Electric's wind division currently has the largest market share of any wind turbine manufacturer in the world, with a coverage of 15.5 percent of the market compared to 14.0 percent, for its nearest competitor Vestas, and 9.5 percent for the third-ranking company, Siemens, of Germany. General Electric is currently known primarily by its long-standing acronym of GE.

GE's involvement in wind turbines is based on its May 2002 acquisition of Enron Wind following the latter company's demise as a result of the 2001 bankruptcy of its parent company, Enron. Enron Wind, in turn, was formed in January 1997 following the acquisition of an even older wind company, the Zond Corporation of California.

Zond was formed in 1980 by inventor James G. P. Dehlsen. Dehlsen was born in Guadalajara, Mexico, to Danish parents. His father was a businessman and engineer who was working at the time for the Southern Pacific Railroad. Dehlsen later told interviewer and author Peter Asmus (*Reaping the Wind: How Mechanical Wizards, Visionaries, and Profiteers Helped Shape Our Energy Future*) that he had literally grown up in a railroad car. He later went on to attend college at the University of San Diego, from which he received a degree in mechanical engineering.

Dehlsen became interested in both business and engineering challenges soon after leaving college, gaining significant success first in the field of stock investments, founding a company that was eventually worth $200 million. He also founded the Triflon corporation, which sold Teflon-based lubricants. Asmus says that Dehlsen "found religion" with regard to wind power while attending a historic 1981 wind power conference at Palm Springs, California. Dehlsen told Asmus that "[b]y the second day [of the conference] I was convinced that wind was the way to go. All the independent studies on wind were tied together at that conference." He decided to use the profits from his previous business endeavors to begin the development of wind farms in Southern California, the first of which he sited at his own home, partially to prove to doubters that the technology was safe and efficient. Zond later began to install turbines at some of the locations that now hold the largest wind farms in California, most notably at Tehachapi Pass. And, when Dehlsen found that existing turbine models did not operate satisfactorily, he decided to start building his own machines, largely based on the turbines he had visited and studied in Denmark.

Zond experienced dramatic early success in the 1980s fervor for wind power development. By February 1985, the company had installed and was operating 1,900 turbines in California, just over a fifth of all the wind turbines in the state. The company eventually installed turbines at all of the major wind farm sites, Tehachapi Pass, Altamont Pass, and San Gorgonio Pass, with a total of 2,400 machines by 1990.

During the 1990s, however, Zond began to experience many of the same problems affecting the wind industry in general in the United States. As a result, it made itself available for purchase and was acquired in January 1997 by the Enron Corporation, a giant energy firm based in Houston, Texas. The company was renamed the Enron Wind Company, which, in its short five-year career, never made a profit for its parent firm. Enron Wind was then sold once again in May 2002, following the disastrous collapse of the parent company in a historic case of mismanagement and bankruptcy. The new owner of the company was GE, which renamed the company GE Wind Energy and made it a subsidiary of the GE Energy division. In 2013, the company reorganized its energy functions. Its wind turbine functions have been reassigned to one of two subdivisions, GE Energy Management and GE Power & Water.

GE currently produces eleven kinds of wind turbines modeled on three basic machines, the 1.5 MW series, 2.5 MW series, and the 4.1–113 offshore wind turbine. Variations of these basic models are available for specialized purposes, such as the 2.85–103 turbine especially designed for Japan. The company also makes and sells ancillary equipment, such as turbine towers and energy storage components that use battery systems to store turbine-produced electricity.

Paul Gipe (1950–)

Gipe has been a spokesperson and advocate for wind energy for almost four decades. He has written seven books on the subject: *Wind Energy: How To Use It* (Stackpole Books, 1983), *Wind Power for Home & Business* (Chelsea Green, 1993), *Wind Energy Comes of Age* (John Wiley & Sons, 1995), *Glossary of Wind Energy Terms* (Folaget Vistoft, 1997), *Wind Energy Basics* (Chelsea Green, 1999), *Wind Power: Renewable Energy for*

Home, Farm, and Business (Chelsea Green, 2004), and *Wind Energy Basics: A Guide to Home- and Community-Scale Wind Energy Systems* (Chelsea Green, 2009). Some of his books have been translated into Spanish, French, and Italian, and *Wind Energy Comes of Age* was selected by the Association of College and Research Libraries as one of its outstanding academic books of 1995.

Gipe has received numerous awards for his contributions to the advancement of wind energy, such as the first Lifetime Achievement Award of the Small Wind Conference in 2013, Individual Leadership Award of the Canadian Wind Energy Association in 2009, the World Wind Energy Award of the World Wind Energy Association in 2008, and the Sierra Club Cup of the Kern-Kaweah Chapter for outstanding service in 2005. He has also been named a "Pioneer in Renewable Energy" by the World Renewable Energy Congress in 1998 and "Man of the Year" by the American Wind Energy Association in 1988. He has spoken at and presented papers to wind organizations in almost every part of the world, including the American Wind Energy Association, Irish Wind Energy Association, American Solar Energy Society, American Society of Civil Engineers, British Wind Energy Association, California Wind Energy Association, European Wind Energy Association, Indian Solar Energy Society, Hannover Messe, New Zealand Wind Energy Association, Australia-New Zealand Solar Energy Society, Asia-Pacific Wind Energy Centre, Nordvestjysk Folkecenter for Vedvarende Energi, and the American Society of Mechanical Engineers.

Paul Gipe was born in Alexandria, Indiana, on October 16, 1950. He told the author of this book that he first became interested in science when his grandparents bought him a "science experiments you can do in your own home" book when he was nine or ten years old. During his high school years, he organized a math and science club that was perhaps most famous (or infamous) for building a rocket-propelled balsawood car that members tested in the school's halls after classes were over for the day.

Gipe said that a defining moment in his life came when he was awarded a National Science Foundation (NSF) fellowship to study field ecology at the University of Montana between his junior and senior years. The opportunity this experience provided him, he said, set him on the path that he continues to follow today, and his interest in wind energy is a manifestation of his background in environmental studies.

After graduating from high school, Gipe enrolled at the General Motors Institute (GMI) in Flint, Michigan, primarily at the urging of his parents, he noted. His plan was to major in mechanical engineering, but he was not very happy at GMI and transferred to Ball State University in Muncie, Indiana, in 1971. Two years later he received his bachelor of science degree from Ball State, having already been recognized as the outstanding student in the Natural Resources Department for 1972–1973. While at Ball State, Gipe was chair of the university student association and was a member of the environmental action committee.

In 1973 Gipe accepted a job with Gannett, Fleming, Corddry, and Carpenter, Inc., an international planning, design, and construction management firm, as an environmental scientist. In this job, he worked on an interdisciplinary team preparing environmental impact statements. He also collected and analyzed meteorological, air quality, and ambient noise data. In 1976, Gipe left Gannett, Fleming, Corddry, and Carpenter to form his own firm, Paul Gipe & Associates, which he operated until 2004. From 1980 to 1984, he also worked as director of the Center for Alternative Resources and from 1984 to 1985 as director of corporate communications for Zond Systems (now GE Wind).

In 2004, Gipe moved to Ontario to take the post of acting executive director at the Ontario Sustainable Energy Associate, where he later became senior policy analyst. In 2009 he assumed a similar role at the Green Energy Act Alliance, an Ontario-based group whose goal it was to make the province a world leader in the development and use of green energy. Since

2003, Gipe has spent a major part of his time working on advanced renewable tariffs for wind energy, payments for energy generated by renewable resources.

Daniel Halladay (1826–1916)

In the 1850s, a water pump salesman by the name of John Burnham was thinking about a way of making his product more appealing to and more useful for farmers. In much of the land west of the Mississippi, pumps were needed to bring groundwater to the surface for personal use, irrigation of crops, and watering of domestic animals. But raising of groundwater was an arduous and time-consuming task, and it occurred to Burnham that a simpler, more efficient, and inexpensive method for raising water was available (and had been for centuries): wind power.

Burnham turned to one of his colleagues who operated a machine shop in his town, Daniel Halladay, for advice in solving this challenge. He told Halladay that if he, Halladay, could design an efficient water mill for lifting water to the Earth's surface, he, Burnham, could sell as many of the wind machines as they could build. Halladay saw no problem in the challenge. He told Burnham, "I can invent a self-regulating wind mill that will be safe from all danger of destruction in violent wind storms: but after I should get it made, I don't know of a single man in all the world who would want one."

And that was almost the scenario that played out. Halladay designed a windmill like one that had never been seen before in that it was, indeed, self-regulating. It automatically turned in the direction of the wind and adjusted its panels so as to take full advantage of the wind power available at any given speed. It also automatically shut down when winds became too strong for the machine to operate safely, although it also operated in even the mildest breeze. In spite of its many beneficial features, Halladay's windmill was of little or no interest to farmers in the East, and it was not until Burnham and Halladay moved their operations to Chicago that the machine finally found its market. In the ensuing decades, the Halladay windmill became one of the most popular devices on American farms, and iconic image of American agriculture.

Daniel Halladay was born in Marlboro, Vermont, on November 24, 1826, to David Daniel Halladay and Nancy M. Carpenter Halladay. When young Daniel was twelve, his family moved to Massachusetts, settling first in Springfield and later Ware. He was educated in the public schools at all three towns, eventually taking an apprenticeship to a machinist for six years. During the latter half of this period, he was foreman at the American Machine Works in Springfield and later at Seth Adams & Company in South Boston. At the end of his period as an apprentice and journeyman, he returned to the American Machine Works, where he was involved in the construction of a revolutionary new caloric (hot air) engine that was eventually displayed at the 1851 World's Fair in London.

Halladay's work on windmills began after his return from London in 1851. He purchased a machine shop in Ellington, Connecticut, which lasted only a short time, and then a similar shop in South Coventry, Connecticut, where he began the manufacture of his new windmills. Lacking the success he had hoped for in his native New England, Halladay eventually moved his manufacturing operations in 1856 to Batavia, Illinois, where his product was much more widely received. In fact, he was so successful with the new business that he was able to sell it at a considerable profit in 1880 and retire to Southern California. He settled in Santa Ana, where he became involved in a variety of business activities. He was appointed president of the Commercial Bank of Santa Ana when it was established in 1882 and remained affiliated with the bank for many years thereafter in one position or another. He was also associated with the Orange County Savings Bank and Bank of Orange. Halladay's wide-ranging interests also led him to affiliations with the fledgling Santa Ana Gas Company and the Santa Ana Orange & Tustin Street Railway.

In the later years of his life, Halladay continued to maintain his interest in a variety of community and cultural activities, including the temperance movement and research on the history of California's Orange County. He died on March 1, 1916, at his home in Santa Ana.

Heron of Alexandria (c. 10–70 CE)

Most histories of wind power mention Heron (also Hero) of Alexandria as being the first person to build a machine that operated with the energy of wind. That machine consisted of a windmill attached to an air pump, which, in turn, was attached to a pipe organ. When a gust of wind appeared, it caused the windmill to turn. The motion of the windmill blades was transferred to the up-and-down motion of the air pump, which forced air into the organ, causing sounds to occur. In retrospect, the device seemed relatively simple to design and operate. One gains perspective on its importance, however, by realizing that no similar wind machine had ever been produced prior to Heron's time.

Relatively little is known with certainty about Heron's birth and early life. Some scholars place his birthdate at about the year 10 CE, although later dates have also been proposed. Perhaps the best biographical information we have about Heron is that we know he must have been alive and working in 62 CE, when he reported observing an important eclipse that took place that year.

Heron's birthplace is also uncertain, although he is generally thought to have been born and raised in Alexandria, where he also spent most of his life. Many scholars believe that he was a teacher at the famous Museum of Alexandria, at least partly because some of his surviving works read as if they are lecture notes rather than didactic texts.

A rather large number of books have been credited to Heron, although it is unclear as to whether he was the actual author of some of them. It is possible that some works attributed to him were actually translations or interpretations of earlier works. Among the books that he is most likely to have produced are *Automata, Barulkos, Belopoiica, Catoptrica, Cheirobalistra, Definitiones, Dioptra, Geometrica, Mechanica, De mensuris, Metrica, Pneumatica,* and *Stereometrica.* The topics of most of these books can be construed from their names (*Geometrica* = geometry), and all deal with either technical or mathematical subjects. *Pneumatica,* again as the title suggests, discusses topics associated with gases, especially a number of Heron's inventions that used air, steam, or other gases for their operation. The first mention of the wind organ appears in this volume.

Much of the written record that remains of Heron's works focuses on simple machines or instructions for the manufacture of practical devices (e.g., military weapons) rather than theoretical formulations. One exception is in the field of mathematics, where Heron developed a number of solutions and formulas. The most famous of these works is now known as Heron's formula, a procedure for calculating the area of a triangle knowing the lengths of its three sides: $A = \sqrt{s(s-a)(s-b)(s-c)}$, where A is the area of the triangle, s is the semiperimeter of the triangle (half the distance around the triangle), and a, b, and c are the lengths of the sides of the triangle.

Neither the date nor the place of Heron's death is known. Conjecture places them at Alexandria in about 70 CE.

William ("Bill") Edward Heronemus (1920–2002)

Heronemus has been called by his biographer "the father of modern wind power" for his important contributions to the development of that technology during the earliest years of its development in the 1970s. As early as 1968, Heronemus had written about the world's forthcoming energy crisis, which he saw developing as supplies of fossil fuels became less readily available and the environmental problems they caused more serious. He warned that

> in the immediate future, we can expect the "energy gap" to result in a series of crises as peak loads are not met. The East Coast will be dependent on foreign sources for most of its oil and gas. The environment will continue to deteriorate in spite of ever-increasing severity of controls. Air pollution, oil spills and thermal pollution are likely to be worse, not better in 1985. In the face of the continuing dilemma: power us. pollution: a third alternative [to nuclear and fossil energy] must be sought. It may be found in the many and varied non-polluting energy sources known to exist in the US or its offshore aggregate. These energy sources, tied together in a national network, could satisfy a significant fraction of our total power needs in the year 2000. (Quoted in: Appreciation: The Life and Work of Bill Heronemus, Wind Engineering Pioneer. http://theheronemusproject.com/THP/library/Stoddard%20The%20Life%20and%20Work%20of%20Bill%20Heronemus.pdf. Accessed on April 25, 2014.)

Heronemus is credited not only with writing and speaking about the potential of wind power to deal with the coming energy crisis but also with directing research and teaching students to deal with the technical issues of making alternative and renewable energy a reality for the coming decades.

William Edward Heronemus was born in Lancaster, Wisconsin, on April 16, 1920. After graduating from high school, Heronemus enrolled at the University of Wisconsin but left after one year to accept an appointment to the U.S. Naval Academy at Annapolis, Maryland. He graduated from the academy in 1941 and received his commission as an ensign in the U.S. Navy, just two weeks after the bombing of Pearl Harbor and the entry by the United States into World War II. He served as a gunnery officer on the destroyer USS *Woodworth* and took part in a number of Pacific campaigns, including the Battle of the Solomon Sea in 1942, for which he was awarded the Bronze Star with the Combat "V." He was later reassigned to the Construction Corps and transferred to the battleship *North Carolina,* where he served as the engineering auxiliary officer. Heronemus was decommissioned from the navy in early 1945, allowing him to continue his studies for the master of science degree at the Massachusetts Institute of Technology (MIT). He earned his degree there in naval architecture and marine engineering. It was at MIT that Heronemus learned about the pioneering work of Palmer C. Putnam and Percy Thomas on wind turbines that was to motivate much of his own research at the University of Massachusetts two decades later.

Still on active duty with the navy, Heronemus was assigned to work at the Portsmouth (New Hampshire) Naval Shipyard, where his research focused on the design of nuclear submarines. He retired from the navy in 1965 with the rank of captain and took a position with the United Aircraft Corporation as engineering manager, where he continued to work on nuclear power systems. Two years later, Heronemus was offered the opportunity of creating a program in ocean engineering at UMass that included study and research on oceanography, marine biology, and geology in addition to his own field of naval architecture. During his first years at UMass, Heronemus began to distance himself from his previous work on and advocacy of nuclear power, drifting toward a greater interest in the role of renewable energy sources for the future. He focused his attention in particular on wind systems, especially those that could be built and operated in offshore environments.

An overview of the kind of wind systems Heronemus had in mind was offered by *The National Geographic* magazine in its December 1975 issue in an article titled "Can We Harness the Wind?" The article described "gigantic wind towers rearing out of a calm ocean, each of them equipped with three 200-foot-diameter wind generators" that, according to Heronemus, "could supply 2 1/2 times the amount of electricity consumed by the New England states last year, [at a] cost . . . 45 percent less than the cost of building nuclear plants with the same capacity" (*The National Geographic* article is available online at http://www.

getsolar.com/blog/the-future-of-wind-energy-in-1975/104/). Heronemus had earlier brought together a number of UMass professors and researchers to present a proposal to the National Science Foundation (NSF) for the development of a Grand Scale Offshore Wind Power System that eventually resulted in the construction of one of the first and largest wind turbines in the United States (if not the world) in 1972.

In 1998, Heronemus took leave from UMass to found Ocean Wind Energy Systems with one of his former students, Forrest "Woody" Stoddard. The company not only brought together much of Heronemus's earlier research and collected his patents on that research, but also began new studies on the development and application of wind systems. Heronemus died in Amherst, Massachusetts, on November 2, 2002, after a long battle with cancer. He had been honored with the chancellor's medal from UMass in 1977 and the Lifetime Achievement Award of the American Wind Energy Association in 1999.

Marcellus Jacobs (1903–1985)

Marcellus ("M. L.") Jacobs and his brother Joe established the Jacobs Wind Electric Company in 1928 to manufacture small wind turbines for use primarily on farms. The business grew out of popular response to a wind system they built on their own farm in eastern Montana to provide electricity for their own use. At first, the Jacobs had relied on a conventional Delco generator to supply electricity for their lights, water pumps, radios, irons, and other appliances. But obtaining the gasoline they needed to operate their generator involved a three-day round trip visit to the nearest town that carried the fuel caused them to think about alternative means of electrical generation. It was that problem that led them to the development of their first windmill-like wind turbine. The Jacobs Company soon gained a wide reputation as a dependable supplier of simple wind machines for home and farm use. The company remains in operation today.

Marcellus Jacobs was born in Cando, North Dakota, in 1903. He was one of eight children in a family that moved from North Dakota to Indiana and then on to Montana. He finished high school and studied electricity in Indiana before completing a six-month course in Kansas City. But he told an interviewer for *Mother Earth News* in 1973 that he always thought of himself as being self-educated. He remembers that he was always interested in electrical circuits and wind and realized early on in his life that he should be able to make a machine that converted the power of the latter into the former.

The Jacobs brothers encountered a host of technical problems in converting the widely popular windmills used for lifting water and grinding grain into electrical generators. Eventually they solved those problems and developed a working system for their own ranch. As neighbors heard about and saw these systems, they asked the Jacobs brothers to build similar system for them, and between 1927 and 1931, they built about two dozen such systems in Montana. In 1931, they finally decided to sell their ranch and commit themselves to the wind turbine industry, incorporating first in North Dakota and moving later to Minneapolis.

The Jacobs Wind Electric Company went through many years of success and tribulation. For many years, it was the leading producer of wind turbines in the United States and the largest such producer of small machines in the world. With the advent of the Rural Electrification Act in 1935, the company faced greater competition than it could handle, and it went out of business. With the revived interest in renewable energy in the 1970s, however, it returned to business and has remained active ever since. Marcellus Jacobs retained his affiliation with the company until his death on July 15, 1985, as the result of an automobile accident in Minneapolis.

It is of interest to note that most of the information currently available about the Jacobs Wind Electric Company refers to the contributions of Marcellus. The company's own Web site mentions his brother, Joseph (Joe), but provides almost no information about him. Also, Marcellus never mentioned Joe in his extended interview with *Mother Earth News*. Yet, according to one commentator in the October/November 1988 issue of The *Home Power* magazine, Joe Marcellus was "the real brains behind the Jacobs wind generator," and the brother who received the first patent on the wind turbine in 1918. Joseph Jacobs, like his brother, was killed in an automobile accident in the early 1960s.

Poul la Cour (1846–1908)

Poul la Cour is one of the most famous names in the history of wind power generation. While a teacher at the Askov Folk High School in west central Jutland (Denmark), he conducted a number of experiments on the use of windmill-like devices to generate electricity from wind power. In his research, La Cour attacked some of the most basic problems related to the functioning of wind turbines, including the creation of methods to collect and store electrical energy produced by turbines, thus evening out the irregularity of their production, as well as devising methods for allowing turbines to operate effectively at a variety of wind speeds.

La Cour was born on April 13, 1846, on a farm near Ebeltoft, Denmark. He grew up on a farm in Djursland owned by his father, an enthusiast for the use of modern farming technology for the improvement of agriculture. He and his parents hoped that Poul could train for the priesthood, but he did poorly in his early school years, earning a failing grade in Greek but an A+ in mathematics. In his search for a new vocation, he followed the suggestion of his older brother Jørgen, who convinced him of the growing importance of the field of meteorology. He enrolled at the University of Copenhagen, where he experienced considerable success, earning his master's degree in 1869 and a silver medal for his thesis on the properties of cloud layers.

Over the next three years, La Cour served a term as a corporal in the Danish Army, taught for a year at the Lyngby Agricultural School, and toured England, Italy, the Netherlands, and Norway. During his stay in the Netherlands, he met and worked with the famous Dutch meteorologist C. H. C. de Buijs Ballot. When he returned from his trip, he assisted in the creation of the new Danish Meteorological Institute in 1872. He served as deputy director there, working to establish weather stations at a variety of locations in Denmark.

During his five-year tenure at the meteorological institute, La Cour worked on a number of his own inventions, one of the most important of which was his tuning fork telegraph. The device was created as a way of sending multiple messages by telegraphy at the same time by splitting those messages into separate components using a tuning fork–like mechanism. Although adopted by the Danish Railroad Company, the system never received wider application because of the consequent success of similar systems developed by a number of American inventors.

In 1878, La Cour was invited to join the teaching staff at the Askov Folk High School as a science teacher. The high school had been established on the principles of Grundtvigianism, a philosophy based on the teachings of the Danish cleric Nikolaj F. S. Grundtvig in the mid-nineteenth century. The philosophy was deeply religious and aimed at helping people live a pure and simple life on Earth. The authorities at the Askov school had decided that they needed to increase their emphasis on the natural sciences, a task that was not particularly easy because of the traditional conflict between religion and modern scientific teachings. La Cour seemed the ideal person to fill that slot, however, since he was both a proficient and highly respected scientific researcher and a committed follower of Grundtvigianism.

One of the projects on which La Cour focused his attention at Askov was the construction of a wind turbine for the generation of electricity. He saw the project as the perfect combination of his scientific skills with the use of natural resources to provide a better life for rural families. In 1891, he obtained a grant from the Danish government for building a windmill-like device that could produce electricity from wind energy. The electrical energy thus generated was used to electrolyze water into hydrogen and oxygen gases as a way of storing the otherwise evanescent wind energy. His demonstration project was a major success, and he was soon involved in the construction of similar machines at a number of rural locations throughout Denmark.

La Cour did not ignore other topics of interest while he was working on his wind turbines. For example, he was curious as to the precise physical properties a wind turbine needed in order to function with maximum efficiency. To answer this question, he built one of the world's first wind tunnels to study the behavior of various types of wind turbines in different wind conditions. He also developed a number of technical variations on the original wind turbine to improve its operation. La Cour died in Askov of a lung infection on April 24, 1908.

La Cour is remembered not only for his development of the wind turbine and his original research in a number of scientific areas but also for his interest in and promotion of the history of science. He wrote two important books on the topic: *Historisk Mathematik* (1881) and *Historisk Fysik* (1896–1901). He was also an ardent supporter of the new emphasis on physical training, as part of the Grundtvigianism philosophy. He saw competitive sports as being not so much as a competitive activity as an opportunity for social interaction and the development of one's physical skills.

Rudd Mayer (1943–2002)

Mayer was an ardent environmentalist with a special interest in the contributions that wind power can make to providing the nation and the world with a safe, clean, renewable energy.

She came to environmental activism somewhat late in life, after having lived a somewhat conventional life that included raising four children. She was born in Illinois and spent most of her life there but made an important decision in the 1990s to move to Boulder, Colorado. When she arrived in Boulder, she said that she seemed to feel that she had "come home" to the open spirit of the mountainous west that she had first visited at the age of ten. At that point, she started to become active in environmental affairs, joining the Land and Water Fund of the Rockies (LAW; now the Western Resources Advocates), an organization with which she was affiliated for ten years until her death in 2002.

Perhaps her signal contribution to the environmental movement was her involvement in a program allowing the state's publicly owned utility, Public Service Company (PSC) of Colorado, to include wind power in its energy offerings. The company was somewhat uncertain as to the willingness of Coloradans to spend a bit more on their energy bills to pay for wind power, which was not yet economically competitive with fossil-fuel-generated electricity. Mayer organized a program to make the new PSC program better known and to encourage customers to participate in the wind energy option. Her Grassroots Campaign for Wind Power carried out an outreach program to not only residential customers of PSC but also small businesses, whose needs are sometimes ignored by large utilities. Mayer's Grassroots volunteers literally went from door to door to contact small businesses and make them aware of the wind energy option available from PSC.

Mayer's efforts were ultimately very successful in promotion of the PSC Windsource Program. She later expanded her efforts to other states, such as Arizona, New Mexico, and Utah, where utilities have adopted similar wind energy initiatives as part of their own energy systems. Her model for the promotion of wind energy options has also been adopted in one form or another by a number of environmental organizations throughout the United States.

Rudd Mayer was born in Washington, D.C. on September 8, 1943, to Dorothy and Rudyard K. Magers. She spent most of her childhood in Evanston, Illinois, where she attended local elementary schools and Evanston High School. She then matriculated at Smith College, in Northampton, Massachusetts, from which she received her bachelor's degree in art history. After graduation, she married her high school boyfriend, Richard Mayer. When the couple returned to Illinois, Rudd concentrated on being a wife and mother to her four children, while working as a copy editor at an advertising agency in the Chicago area.

After her divorce, Mayer decided to move to Boulder, where she began a completely new and different life. Although she had always been interested in and involved with a variety of outdoor sports, her work on the behalf of wind energy was a new adventure, to which she devoted all her energies. Her relatively short career as an environmental activist came to an unexpected end when she died of heart failure said to have been associated with an unspecified childhood illness.

Mayer's work in the field of wind energy has been recognized in a number of ways. In 2001, she received the Green Power Pilot Leadership Award sponsored by the Environmental Protection Agency (EPA) and the Department of Energy (DOE), an award that was renamed in her honor the Rudd Mayer Green Power Pilot Award a year later. Previously she had also received the National Award for Sustainability from the President's Council on Sustainable Development and Renew America in 1999 and the American Wind Energy Association's 1999 Wind Energy Advocacy Award. In 2002, she was also given the 2002 EPA Climate Protection Award. Also, in 2002, the Thorne Ecological Institute established the Rudd Fund to provide educational experiences for children and young adults about renewable energy resources. The organization Women of Wind Energy (WoWE) has also created the Rudd Mayer Memorial Fellowships to support the education and work of women in the field of wind energy. (See the WoWE entry in this chapter for more information.)

Mexican Wind Energy Association

Ave Jaime Balmes No. 11 L 130 F
Col. Los Morales Polanco
Mexico, DF 11510
Mexico
Phone: +52 55 5395-9559
Email: direccion@amdee.org
URL: http://www.amdee.org/Inicio.html

Many countries and regions around the world now have national wind energy organizations, created to support and encourage the development of wind power projects in those nations and regions. A few examples of such organizations are the Bulgarian Wind Power Association, East African Wind Energy Association Hungarian Wind Energy Scientific Association, Macedonian Wind Energy Association, Societe Algerienne de l'Energie Solaire (Algeria), Taiwan Wind Energy Association, Turkish Wind Energy Association, and Vindkraftforeningen (Finland). Many of these organizations have common forms of governance and participation and common programs and events. This entry for the Mexican Wind Energy Association is offered as an example of the character and function of these organizations.

The Asociacion Mexicana de Energia Eolica (AMDEE; Mexican Wind Energy Association) was founded in 2005 to promote and encourage the development of wind energy projects in Mexico. The association has four categories of membership: wind energy developers, equipment manufacturers, service providers, and honorary members. Examples of current members of each category are as follows:

Developers: ABB Mexico, Alatec, Industrial Wiring, Carbon Solutions from Mexico, Mexico Wind Developers, Enel

Green Power Mexico, GE International Mexico, Renewable Iberia, and Siemens.

Equipment manufacturers: Alstom Mexicana, Industry Boom, Industrial Power, and Vestas Mexico.

Service providers: Fresh Energy Consulting, Green Momentum, Windward Group, RSA Insurance, and Walbridge of Mexico.

Honorary members: Energy, Technology, and Education (ENTE, S.C.) and Institute of Electrical Research.

The makeup of other national and regional wind organizations tends to be very similar to this pattern.

The work of AMDEE is divided into about a half-dozen major projects: regulatory issues, funding, development projects in Mexico, land regularization, climate change, and land use issues. The organization sponsors a national conference on wind energy in Mexico, and other meetings on specific topics of interest to wind energy developers. These sessions deal not only with general problems and issues of wind power generation but also with specific concerns for developers and consumers of wind energy in Mexico.

The organization's Web site is a good source of general information about wind power with one section ("The Wind") providing basic information about how wind energy works, small and large applications of wind power, the status of wind energy in the world, and myths and realities about wind power. The Web site also offers a useful section on wind energy news, with an obvious focus on events taking place in Mexico.

The Web site is particularly helpful because its basic language is Spanish. However, it also provides ready translations of the Web site material into forty-six other languages, ranging from English and French to Bulgarian, Icelandic, and Tagalog.

National Renewable Energy Laboratory

15013 Denver West Parkway
Golden, CO 80401

Phone: (303) 275-3000
Washington, D.C. Office
901 D Street, SW, Suite 930
Washington, DC 20024-2157
Phone: (202) 488-2200
Email: http://www.nrel.gov/webmaster.html
URL: http://www.nrel.gov/

The National Renewable Energy Laboratory (NREL) was created by a provision of the Solar Energy Research Development and Demonstration Act of 1974. It was originally called the Solar Energy Research Institute (SERI) and began operations in 1977. Over the next fifteen years, the laboratory expanded its areas of research and was named a national laboratory in 1991, and its name was changed to its present form. The laboratory is operated for the U.S. Department of Energy under contract by the Alliance for Sustainable Energy, LLC, a partnership of Battelle and MRIGlobal.

In its earliest years, SERI and NREL were well funded by presidential administrations that were supportive of efforts to increase the nation's research on renewable energy. Under President Ronald Reagan, who favored greater emphasis on traditional fossil fuels, however, the laboratory's budget was cut by about 90 percent. In recent years, the laboratory has once more received more generous support from the federal government, although a number of critics are still working to reduce funding for the agency or to eliminate it altogether. NREL's projected research budget for fiscal year 2014 is $271 million, a significant decrease from its peak funding of $536.5 million in 2010, and down from $371.6 million in 2013.

The work of NREL is divided into five major categories: energy analysis, science and technology, technology transfer, technology development, and energy systems integration. The results of the laboratory's most recent work in the first of these categories are summarized in a publication, "Realizing a Clean Energy Future," in which the laboratory describes its research on ways in which the nation and the world can smoothly undergo a transformation from a carbon-based economy to one based on a variety of renewable energy sources. The science and technology sector focuses on basic and applied research on the major forms of renewable energy: photovoltaics, wind, solar power, geothermal energy, water power, biomass, and hydrogen and fuel cells. It also pursues research on a number of related topics, such as electricity integration, building efficiency, vehicles and fuels, biosciences, and computational science.

The area of technology transfer deals with the problems and issues involved in converting scientific and technological developments into forms in which they can actually be used in the everyday world. It answers questions about agreements for commercializing technology, licensing procedures, technology partnerships, and the development of facilities. Technology deployment is a procedure by which NREL employees and others work to find ways to introduce useful new technologies into the marketplace. Some of the issues with which the laboratory deals are using technology for disaster preparation and recovery, project development, overcoming barriers in the marketplace, and assistance with technical problems. Energy systems integration involves one of the most challenging problems facing the development and use of renewable energy systems: how to integrate these new technologies with a complex existing electrical grid. Some of the challenges that have to be faced include the greater use of electric vehicles, the incorporation of photovoltaics into existing systems, and the home use of wind turbines.

In addition to this range of activities, NREL provides educational opportunities to professional in the field and other stakeholders in renewable energy development. These opportunities most often take the form of seminars and workshops on topics such as Stability Analysis of Microgrid Systems, National Energy and Transportation Infrastructure Design, the Power System of the Future: Innovations, Trends, and Signposts, Power Grid Data Management and Analysis, Electrical Visualization and Operations, Integrated Deployment, Hydrogen and Fuel Cells, and from the Building to the Grid: An Energy Revolution and Modeling Challenge.

National Wind Coordinating Collaborative

c/o American Wind Wildlife Institute
1110 Vermont Ave., NW, Suite 950
Washington, DC 20005
Phone: (202) 656-3303
Email: info@awwi.org
URL: http://nationalwind.org/

The National Wind Coordinating Committee (NWCC) was formed in 1994 at a meeting of the National Renewable Energy Laboratory (NREL), U.S. Department of Energy (DOE), American Wind Energy Association (AWEA), National Audubon Society, Electric Power Research Institute (EPRI), and Union of Concerned Scientists (UCS). The organizing conference was called the National Avian-Power Planning Meeting. The meeting was called to identify technical questions relating to the interaction of wind turbines and avian species. The basic philosophy behind the meeting was that organizations concerned with the potential threat posed to birds by wind farms should have the opportunity to get together in a neutral setting and agree as to the issues that are involved in this setting and then collect, summarize, and, if necessary, add to the scientific research dealing with this issue.

The NWCC is co-funded by the Wind and Water Technologies Program of the U.S. Department of Energy's National Renewable Energy Laboratory and the American Wind Wildlife Institute (AWWI), with which NWCC is affiliated. Membership in NWCC is open to any and all parties interested in wind–wildlife issues who are willing to agree to the general principles under which the organization operates. Those guiding principles were first developed at the 1994 organizing meeting and have since been updated on a number of occasions. The current statement of operating principles is available online at http://nationalwind.org/wp-content/uploads/assets/wildlife/NWCC_WLWG_Participation_Guidelines__08-17-12_.pdf.

Some of the topics with which NWCC has dealt since its creation include conceptual models and/or frameworks of the principal causes of avian mortality at wind plants; a common set of metrics that researchers can use to characterize the potential or existing impact from a wind development; short- and long-term impact studies; risk assessment; avian behaviors and turbine properties that lead to collisions; local topographic and geographic influences on bird migration; effects of turbines on bats; displacement effects of wind developments on grassland birds; effects of wind development on nesting behaviors; habitat displacement; population impacts; on-site impact reduction techniques and offsite mitigation; curtailment studies; cumulative and landscape-scale impacts; and adaptive management policies for dealing with avian-wind power interactions.

NWCC activities are currently divided into five major categories: transmission, siting, green power marketing and credit trading, environmental benefits and costs, and distributed generation. A variety of activities have been carried out within each of the general categories, most of which consist of meetings, research projects, seminars and webinars, presentation, briefings, and publications. In addition, NWCC maintains an archive of publications dealing with wind–wildlife issues, such as wind energy issues papers on the benefits of wind energy, wind energy environmental issues, siting issues for wind power plants, wind energy resources, and wind energy costs; issue forms on topics such as New England wind issues–wind partners coordination and wind energy economic developments; research reports on the interaction of wind turbines with birds and bats; and a collection of monthly updates from the organization.

The NWCC Web site also provides very useful information about national, state, and local resources on wind energy and wind–wildlife interactions, with links to organizations such as the U.S. Department of Energy, Management, U.S. Fish and Wildlife Service, Association of Fish and Wildlife Agencies (AFWA), Association of Fish and Wildlife Agencies, the Western Association of Fish and Wildlife Agencies (WAFWA), Western Governors' Association (WGA), American Wind Energy Association (AWEA), Bats and Wind Energy Cooperative (BWEC), Great Lakes Wind Collaborative (GLWC), and Partners in Flight Logo.

NWCC sponsors and co-sponsors two major meetings on a regular basis (usually biennially): the Wind Wildlife Research Meeting and the National Avian-Wind Power Planning Meeting. These meetings bring together academics, researchers, conservation scientists, consultants, federal and state officials, representatives of nongovernmental organizations, and wind industry professionals. The NWCC Web site also provides an update on important news about topics related to wind–wildlife issues.

Office of Energy Efficiency & Renewable Energy

Forrestal Building
1000 Independence Ave., SW
Washington, DC 20585
Phone: None provided
Email: None provided
URL: http://energy.gov/eere/
office-energy-efficiency-renewable-energy

The parentage of the Office of Energy Efficiency & Renewable Energy (EERE) dates to 1973 when President Richard Nixon declared his Project Independence, a program designed to make the United States less reliant on fossil fuels then purchased primarily for the oil-exporting states of the Middle East. He added a component to the existing Department of the Interior's program for research on coal, oil, and natural gas for similar studies of alternative fuels, although very little was actually done through that program.

In 1975, President Gerald Ford established the Energy Research and Development Administration (ERDA) as a way of placing greater emphasis on energy research issues, including renewable energy sources. Once again, relatively little was done on the development of alternative fuels. In 1977, President Jimmy Carter created the Department of Energy, into which was absorbed the ERDA. In 2001, the agency was re-designated under its current name.

The EERE had a budget of $1.8 billion during fiscal year 2011, of which $692 million went to research on renewable fuels. The largest share of that money went to research on solar energy ($260 million), followed by research on biomass ($184 million), hydrogen and fuel cells ($100 million), wind ($80 million), geothermal ($38 million), and water ($30 million). The major part of EERE's budget was spent on energy efficient projects, such as research on more energy-efficient buildings, industrial operations, and vehicles.

EERE activities are organized into about a half-dozen major categories: energy basics; states and local communities; technical assistance; energy analysis; jobs, education, and training; publications; and funding.

Energy basics is primarily a Web-based service that provides general information on almost every conceivable topic in the area of energy efficiency and renewable energy, such as solar, wind, geothermal, and biomass sources of energy; industrial energy efficiency; passive solar building design; water heating; lighting and daylighting; and space heating and cooling.

States and local communities makes available information about energy efficiency and renewable energy programs and activities in all fifty U.S. states. It also describes a variety of federal and state programs on specific aspects of these topics, such as Advanced Manufacturing Office, Better Buildings Neighborhood Program, Building America, Building Energy Codes Program State Technical Assistance, Clean Cities, Clean Energy Application Centers, Database of State Incentives for Renewables and Efficiency, Geothermal Technologies Office, Green Power Network, Integrated Deployment State and Territory Projects, Standard Energy Efficiency Data Platform,

Tribal Energy Program, State and Local Energy Efficiency Action Network, SunShot Initiative's Ask an Expert, Technical Assistance Program Solution Center, Weatherization & Intergovernmental Programs Office, and Wind Powering America.

Technical assistance offers individuals, governmental agencies, and businesses aid in dealing with a number of energy efficiency and renewable energy issues through agencies such as the State and Local Solution Center, Weatherization Assistance Program Technical Assistance Center, Tribal Energy Technical Assistance, FEMP Technical and Project Assistance Officer, Geothermal Maps, Geothermal Software and Data, Building America Solution Center, Building Energy Codes Program, Building Energy Software Tools Directory, Building Energy Codes Program State Technical Assistance, Fuel Cell Technologies Education, Combined Heat and Power Technical Assistance Partnerships, Technical Assistance for Manufacturers, Solar Outreach Partnership, Solar Technical Assistance Team, Alternative Fuels Data Center, Clean Cities Technical Assistance, Hydropower Resource Assessment and Characterization, Marine and Hydrokinetic Resource Assessment and Characterization, and Wind Resource Maps and Data.

Energy analysis is an EERE Web site designed to help energy experts and policy makers get access to four major resources in decision making about efficiency and energy resource topics: data resources, market intelligence, energy systems analysis, and portfolio impacts analysis.

Jobs, education, and training is a resource for individuals wanting to learn more about education for and careers in energy efficiency and renewable energy resources.

Publications is a very large collection of books, articles, brochures, pamphlets, Web pages, and other resources with information on energy efficiency and renewable energy.

Funding is one of EERE's most important functions since it has significant amounts of money available for the support of research and development of and demonstration projects on energy efficiency and renewable energy. The agency's Web site contains a page with detailed information about every aspect of the application and production features of funded projects.

EERE is not only a large and complex agency in and of itself, but it also has connections with a great many other federal and state organizations in the field of energy efficiency and renewable energy. Its web page called Offices (http://energy.gov/offices) is a good starting point for further research on these topics.

Small Wind Certification Council

56 Clifton Country Rd.
Suite 202
Clifton Park, NY 12065
Phone: (518) 213-9440
Fax: (518) 899-1092
Email: info@smallwindcertification.org
URL: http://smallwindcertification.org/

Until relatively recently, a person or organization wanting to purchase and use a small wind turbine had to rely largely on the manufacturer of that device for confirmation that it would perform as advertised; no independent organization existed to test and certify the capability of small wind turbines. In 2006, a group of individuals and governmental agencies met to explore the possibility of creating such an organization for the review and certification of wind turbine models. That informal group created the Small Wind Certification Committee Working Group to bring to reality such a plan. The Working Group was made up of more than sixty individuals and organizations, including the major small wind turbine applicants, representatives from a number of U.S. states and Canada, academic institutions, and key individuals in the field. Funding for the Working Group's activities was provided by the Canadian

Wind Energy Association (with funding from Natural Resources Canada), Casper College (Wyoming), Energy Trust of Oregon, Iowa Energy Center, National Renewable Energy Laboratory, Nevada State Energy Office, and the Wisconsin Department of Administration.

The Working Group spent much of the next two years carrying out background research and determining the most appropriate structure and function for a certification organization. In 2008, they incorporated such an organization as the Small Wind Certification Council, elected a board of directors, and hired a staff. The major goals adopted by the new organization were:

- Supporting the use of small wind turbines in North America and internationally;
- Fostering the exchange and dissemination of information concerning turbine energy and sound level performance; and
- Supporting and fostering appropriate government regulations and legislation related to wind technology issues.

The process of turbine certification involves essentially the testing of a submitted turbine model and determining if that turbine conforms to the standards for such machines established by the American Wind Energy Association (AWEA), known as standard AWEA 9.1–2009, Small Wind Turbine Performance and Safety Standard. Turbines that meet this standard are provided with a label that clearly lists the machine's fundamental properties: rated annual energy production (in kWh/year), rated sound level (in dB[A]), and rated power (in kW).

Certification is a desirable and sometimes essential feature in the process of buying, installing, and using a wind turbine. Electric utilities must know, for example, that distributed turbines that are connected to the electrical grid meet the basic standards claimed for them by their owners. Also, state, federal, and independent funding organizations may require that wind turbines be certified in order to receive various types of funding and tax incentive benefits.

The SWCC Web site is the main database for information about the certification process and about wind turbines that have been tested and approved by the organization. As of late 2014, eight small wind turbines (with "small" being defined as a turbine with a capacity rating of 100 kW or less) from six companies had been tested and approved, whereas one medium turbine (defined as a turbine with a swept area greater than 200 square meters) had received certification. In addition, another nineteen from fourteen companies' turbines were at one stage or another in the testing and approval process.

In addition to the documents available on the organization's Web site, information on a number of features characteristic of small wind turbines is provided, such as consumer resources, news and events about small wind turbines, and a listing of standards and relevant documents about certification.

Utility Variable Generation Integration Group

PO Box 2787
Reston, VA 20195
Phone: (865) 218-4600, Ext. 6141
Fax: (865) 218-8998
Email: http://variablegen.org/contact/
URL: http://variablegen.org/

The Utility Variable Generation Integration Group (UVIG) was formed in 1989 by a group of energy companies and related organizations for the purpose of offering a forum for the analysis of wind and solar technology for utility applications and for serving as a credible source of information to decision- and policy-making groups and the general public on the current status and potential uses of wind and solar energy. The organization was previously known as the Utility Wind Interest Group, Inc., and the Utility Wind Integration Group.

UVIG's mission is to promote the use of good engineering and operational practices in finding ways to provide utilities with greater access to electrical power developed through solar and wind facilities. In achieving this object, the organization works primarily with the U.S. Department of Energy, the National Renewable Energy Laboratory, and research groups from public utilities. As of late 2014, the organization had more than 175 members from the United States, Canada, Europe, and Asia. These members fall into two general groups: corporate members and associate members. Corporate members include investor-owned, public power, and rural electric cooperative utilities, and transmission system operators such as Alaska Village Electric Cooperative, Duke Energy, Fire Island Wind, Hydro Quebec, Idaho Winds, Nebraska Public Power District, Pacific Gas and Electric, Southwest Power Pool, and UpWind Solutions. Associate members include academic and governmental institutions, and energy consultants, such as California Energy Commission, Evergreen Renewable Consulting, Hawaii Natural Energy Institute, Iowa State University, Mitchell Technical Institute, University of Michigan, Wind Utility Consulting, and WindLogics.

A major function of UVIG involves educational meetings of all kinds for its members, including workshops, presentations, and webinars. Topics in the area of wind power of some recent meetings have been "A Mile High View of Wind Integration" (fall 2008 technical workshop in Denver), Distributed Wind/Solar Interconnection Workshop (2013 workshop in Golden, Colorado), "20% Electricity from Wind" (presentation), "Community Wind across America" (webinar), and "Wind Turbine Maintenance Programs" (webinar).

Much of the work done by the organization is carried out through user groups that focus on specific issues of its overall mission. Some of the current user groups are those on variable generation plant modeling and interconnection, operating impact and integration, distributed generation, market operation and transmission planning, and wind/solar plant operations and maintenance. The organization also arranges for peer-to-peer networking on problems and issues of common interest, as well as transfer of technological developments among members.

An important perk of membership is access to all sections of the organization's Web site, which carries proceedings from previous meetings and an extensive list of resources of interest and usefulness to members at all levels of the association. Members also have access to the organization's monthly newsletter, *Connected,* which carries news of the industry and about individual members.

One of the most valuable benefits of membership in the organization is access to its online service, DG Toolbox (DG = distributed generation), which provides information on a host of tools and technology dealing with wind-generated electricity. Some of the resources available on the toolbox are a FERC and flicker screening tool, feeder simulator, program for economics analysis, distributed wind interconnection guide, distributed wind modeling guide, and turbine monitoring report.

An extensive list of reports and other papers is also available on the UVIG Web site classified into categories such as summaries of international studies and experience, NERC (North American Electric Reliability Corporation) Integration of Variable Generation Task Force and other special reports, U.S. regional and state studies, Canadian studies, and European studies. Other individual reports that are available are wind integration state-of-the-art summary, National Conference of State Legislatures wind integration primer, EPRI wind integration impacts summary, and Integrating Wind and Solar Energy in the U.S. Bulk Power System: Lessons from Regional Integration Studies.

Vestas Wind Systems A/S

Hedeager 44
8200 Aarhus N

Denmark
Phone: +45 97 30 00 00
Email: mizar@vestas.com
URL: http://vestas.com/en

Vestas was founded in 1928 by Danish businessman Peder Hansen. Hansen was probably the best-known member of the Hansen family, whose work led to the formation of one of the world's largest wind turbine companies in existence today, Vestas. Hansen's father, Hand Smith Hansen, was a blacksmith who established his own business in the small village of Lem, Denmark, at the age of twenty-two in 1898. H. S. Hansen quickly developed a reputation for his skill in dealing with a variety of inventing and manufacturing challenges, as well as business acumen and general trustworthiness. In 1928, with his son Peder, H. S. Hansen began manufacturing steel window frames, a business that was so successful that he formed a new company, Dansk Staalvindue Industri (DSI) for the manufacture of steel window frames for large commercial and industrial buildings.

DSI suffered a significant setback with the outbreak of World War II, largely because supplies of steel essentially disappeared. After the war, however, Peder Hansen, with a number of his friends and his father, created a new company, Vestjysk Stålteknik A/S, focusing on the production of household appliances, such as mixers and kitchen scales. Before long, the company began to expand its products to include larger objects, such as truck equipment and agricultural machinery. (The company's name was shortened in 1945 to its present title, Vestas, Vestjysk Stålteknik A/S, because of the problems non-Danes had with its original name.)

In 1959, Peder Hansen bought out his partners in order to set a new direction for the company along the lines of his own interest. His enthusiasm for the new opportunity was dampened somewhat, however, when the company's facilities burned down only a year later, in 1960. The apparent setback proved only to be a challenge to Hansen, however, as he threw himself into building the company back up and by the end of 1960, Vestas actually showed an increased profit over the preceding year, with about 100 more employees than had previously worked for the company.

By the end of the decade, Vestas had vastly changed its production mission, focusing on large equipment, such as hydraulic cranes for small trucks, largely for export. As the 1970s dawned, however, the company began to think about yet another addition to its production portfolio, wind turbines. Its first experiments were conducted with vertical-axis wind turbines. That research was carried out largely in secret for fear of the ridicule that might come when the public learned that Vestas was experimenting with giant "eggbeaters." Before long, the company had decided that the future lay with more traditional and more conventional horizontal axis turbines, a product that it eventually began to manufacture in quantity and for which it is best known today.

Although Vestas experienced success with its new venture initially, it, like most of the rest of the wind industry, saw a drop-off of interest in wind turbines beginning in the mid-1980s. In October 1986, the Danish government dramatically reduced its tax credits for the production of wind power, and Vestas began to consider the possibility of bankruptcy. The company responded by deciding to abandon all products other than its wind turbines. The decision proved to be a wise one as modifications in its turbine technology suddenly made the produce widely popular. Vestas actually turned the corner in a business sense in 1990 when it received an order for 342 wind turbines for the Sky River wind farm in California. A year later, the company had delivered its 1,000th wind turbine and largely established its place as a worldwide leader of the wind turbine industry.

By the end of 2012, Vestas had installed more than 51,000 wind turbines with a total rated capacity of 50,000 megawatts in seventy-three countries. In 2013 it announced a new business plan designed to even further extend and expand its wind services around the world. The company currently offers nine different wind turbine models with rated capacities of 1.8, 2.0, 2.6, 3.0, and 3.3 megawatts, meeting the needs of virtually every type of commercial wind facility currently in operation.

Wind Energy Center

Department of Mechanical and Industrial Engineering
University of Massachusetts Amherst
160 Governors Drive
Amherst, MA 01003
Phone: (413) 545-4359
Fax: (413) 545-1027
Email: wec@umass.edu
URL: http://www.umass.edu/windenergy

The Wind Energy Center (WEC) at the University of Massachusetts Amherst (UMass) had its beginning in 1972 when the university received a grant from the National Science Foundation (NSF) to construct an experimental wind turbine on its campus. The project was under the direction of Professor William E. Heronemus, a pioneer in the study of wind energy in the United States. WF-1 (for Wind Furnace 1) was part of a larger UMass project that included the Solar Habitat that aimed at determining the effectiveness and efficiency of a variety of nonconventional energy practices.

The original wind turbine was known as WF-1, had a three-blade design, and was capable of producing 25 kW of energy. By modern standards, that energy output is modest indeed, although at the time of its construction, WF-1 was the most powerful wind turbine in the United States and, perhaps, the world. The construction of WF-1, according to one historian of wind power, marked the beginning of the modern wind power movement. The original WF-1 wind turbine is now on display at the Smithsonian Institution in Washington, D.C.

WEC currently offers a variety of courses and degree and certificate programs in wind energy. Available courses include engineering of windpower systems, wind turbine design, and offshore wind energy engineering, which are offered as part of a bachelor's degree in mechanical and industrial engineering (MIE), a master's degree with a concentration in wind energy, a doctorate in MIE, or a graduate certificate in wind energy. Some of the current research projects being undertaken by graduate students at UMass include wind turbine sound issues, offshore wind site ranking methodology, meteorological and oceanographic study for the Hull offshore wind project, unsteady blade-element momentum theory, wind turbine blade testing, and reliability-based design for offshore wind turbine foundations. Other research topics being pursued by faculty members and adjunct researchers include wind turbine aerodynamics, wind turbine dynamics and controls, external conditions and remote sensing, blade materials and structural design, condition monitoring, turbine monitoring and control, wind resource assessment, and wind diesel-hybrid systems.

In addition to its on-campus programs, UMass offers information about a variety of issues related to wind energy, including information for individuals and communities about the construction of wind turbine systems. The center also provides basic information on wind energy itself with a textbook (*Wind Energy Explained,* 2nd ed., Wiley, 2010), instructional presentations for general audiences (Wind 101 and Small Wind), research publications by faculty and affiliated researchers, fact sheets, guidelines, and other publications on many aspects of wind energy. The center's Web site also provides a number of very useful links to other sites with information about various aspects of wind energy, such as general information on wind power; wind power organizations on the Web; utility and power producer organizations; electricity in Massachusetts and New England; education and research institutions; wind technology; electrical interconnect, transmission, utilities; climate change and weather; planning, policy, and impacts of wind power; and the wind industry.

Wind on the Wires

570 Asbury St., Suite 201
St. Paul, MN 55104
Phone: (651) 644–3400
Email: info@windonthewires.org
URL: http://windonthewires.org

Wind on the Wires (WOW) is a consortium of business, economic development, and community groups; utility representatives; clean energy advocates; and wind energy experts from the states of Illinois, Indiana, Iowa, Michigan, Minnesota, North Dakota, South Dakota, and Wisconsin. The organization was formed in 2001 for the purpose of creating a "road to market" for wind energy, that is, for solving the technological problems involved in transferring the electrical power produced by wind turbines to existing electrical grids. Some current members of the organization are APEX Clean Energy, the American Wind Energy Association, BP, Enel Green Power, Fresh Energy, National Resources Defense Council, Union of Concerned Scientists, Vestas, and WindLogics.

Dealing with this problem involves three subsidiary issues: planning for transmission systems that can accomplish this objective; developing equitable rules for the use of the grid to achieve this end; and educating people who operate and control the grid with regard to the role that wind energy can play in efficient operation of the grid.

These three goals essentially establish the WOW program, which consists of three divisions: technical, regulatory, and educational and outreach. The technical aspect of the organization's activities involves work such as taking part in transmission studies, participating in planning of cost allocations for new facilities, developing market rules that are appropriate for integration of wind power into the grid, and studying existing grid systems to make possible the more efficient integration of energy from renewable sources.

Examples of the regulatory activities undertaken by WOW include participation in federal and state regulatory proceedings to ensure a smooth integration of wind power into grid systems, development of public policies that favor the expanded use of renewable energy in the Midwest, and working on the development and production of a variety of governmental documents dealing with the integration of wind power with the electrical grid. In conjunction with its technical and regulatory activities, WOW has developed an educational and outreach programs for legislators, regulators, consumers, producers, and other stakeholders in the wind energy community.

In its relatively short history, WOW has recorded a number of accomplishments, such as:

- Gaining the support of state regulators for cost allocation plans for seventeen new transmission projects;
- Working with stakeholders to develop favorable siting policies and laws for siting decisions;
- Reaching compromises in the state of Wisconsin for the siting of specific new wind farms;
- Gaining approval from the Minnesota Public Utilities Commission for the creation of new transmission lines in southwest Minnesota;
- Assisting in the development of a 10 percent Renewable Energy Standard in the state of Wisconsin; and
- Advocating for Missouri's Renewable Energy Standards to be met with in-state or imported wind energy.

The WOW web page is a valuable source of information about a variety of topics related to its mission. Its regular News Feed section, for example, consolidates and presents articles, reports, opinion pieces, and other items from wind-related organizations as well as the general media. The Web site also contains a glossary of wind power terms, a resource library, an image gallery, a collection of press releases, links to important external resources (such as the U.S. Department of Energy), and a member toolkit and newsletter (both available to members only).

Windustry

201 Ridgewood Ave.
Minneapolis, MN 55403
Phone: (612) 200-0331
Toll free: (888) 818-0936
Email: info@windustry.org
URL: http://www.windustry.org/

Windustry began its activities in the mid-1990s as a project of the Sustainable Resources Center, a nonprofit organization in Minneapolis, Minnesota, that was being funded by the Legislative Commission on Minnesota Resources. The organization's mission is to promote sustainable energy solutions and to help local communities, businesses, and individuals develop their own energy assets. Windustry works through outreach, educational, and advocacy programs to advance its objectives. In 1999, the organization decided to extend its efforts beyond the state of Minnesota and eventually (2003) became a 501(c)(3) organization partnered with the Institute for Agriculture and Trade Policy (IATP), an organization that works to promote sustainable farms at all levels, from the local to the worldwide.

Windustry held its first conference in 1997 on the topic "Harvesting the Wind." The conference was aimed at providing representatives from local communities and owners of small farms with information about the role that wind energy could play in their lives and businesses. In the same year, the organization produced the first of its educational posters called "Milk This," for the purpose of introducing the benefits of wind power to consumers. In 1999, Windustry launched its web page, which still remains one of its most valuable contributions to the advancement of interest in and information about wind power. The site claims to receive more than a thousand unique visitors each day now. Some of the other achievements that the organization lays claim to are:

- Participation in the Wind Powering America initiative sponsored by the U.S. Department of Education and the National Renewable Energy Laboratory.
- Development of wind energy easement and land use policies and practices for the siting of wind turbines on private property.
- Lobbying and advocacy for the 2002 U.S. Farm Bill, the first farm bill to encourage the use of wind energy for agricultural settings.
- Creation of a wind hotline to provide information and advice primarily to rural individuals and businesses about the use of wind power in their setting.
- Development of a wind integration system in southwest Minnesota to take advantage of buy-in requirements by utility companies of distributed wind energy.
- Production of a booklet called *Harvest the Wind* for the Illinois Institute of Rural Affairs at Western Illinois University.
- Hosting the first luncheon for women involved in the wind industry (Women of Wind Energy; WoWE) in 2004.
- Sponsoring the first Community Wind Energy conference for landowners and communities, a meeting that was to be repeated annually in the following years.
- Establishing in 2009 the Landowner Wind Energy Association Resource Center to provide guidance for landowners in the establishment of successful wind farms on their properties.
- Producing in 2011 the Small Wind Guide, a publication providing basic information for consumers about small wind electric systems.
- Developing a small wind installer training curriculum.

The popularity of the organization's web page can be explained by the range of topics on which it provides information and advice. Its section on wind basics provides an extensive and intensive review of the science and technology of wind energy production, along with a consideration of some of the related problems associated with the installation of wind turbines. Other topics covered on the web page include information on leasing one's land to a wind developer, a community wind "toolbox" that includes information and tools needed for the development of distributed wind systems, a community wind map that shows installed wind facilities in the United States, suggestions for planning a small wind facility, the small wind installer curriculum mentioned earlier, and a list of resources for those who are interested in developing small wind projects.

Another section of the Web site contains a variety of resources helpful to an individual or organization interested in developing a small wind facility. These resources include case studies of successful small wind projects; a list of companies providing materials and services for wind facilities; a collection of documents dealing with all aspects of planning, constructing, and operating a small wind system; a glossary of wind energy terms; a summary of policies past and existing dealing with wind facility development; research on wind energy; and webinars on small wind energy of interest to the prospective producer.

Women of Wind Energy

Kristen Graf, Executive Director
155 Water Street
Brooklyn, NY 11201
Phone: (718) 210-3666
Email: http://www.womenofwindenergy.org/contact-wowe.html
URL: http://www.womenofwindenergy.org/

Women of Wind Energy (WoWE; pronounced WOW-ee) was founded in the spring of 2005 by three women who were employed in the wind industry: Lisa Daniels, Trudy Forsyth, and Mia Devine. (The organization was originally called Women of the Wind but was changed when it was learned that this name was already taken by a group of motorcycle enthusiasts.) The founders created the organization as a way for women working in the industry to share their common experiences, problems, issues, and expectations. The original organization had grown to a membership of more than 1,000 women by the end of 2014. WoWE held its first informal meeting, a luncheon at the 2005 WINDPOWER conference in Denver, which about 120 women attended.

Perhaps the most important event at the luncheon was the announcement of the first Rudd Mayer Memorial Fellowship. The fellowship was created to honor one of the pioneer women in the field of wind energy, Rudd Mayer, who died in 2002. The fellowship provides financial assistance to women in the field for attendance at the annual WINDPOWER conference and other wind energy meetings, for becoming familiar with state-of-the-art wind technology, for meeting their peers in the field, and for connecting with potential employers in wind energy. Six inaugural fellows were announced at the Denver luncheon. Today, the Rudd Mayer fellowship continues to be one of the signature features of the work carried out by WoWE.

In addition to the Mayer fellowship, WoWE annually recognizes two leading women in the field, one an established leader for the Woman of the Year Award, and another a relative newcomer, who receives the Rising Star Award. The organization also provides other forms of assistance to women working in wind energy, including the opportunity to participate in chapters throughout North America, mentoring opportunities both in person and online, webinars on topics of interest to members, and a variety of meetings such as the annual luncheon held at WINDPOWER and the annual leadership form. (An interactive map of WoWE chapters is available at http://www.womenofwindenergy.org/find-wowe-chapter.html#mapAnchor.)

For both members and general readers, one of the most interesting parts of the WoWE program is its feature "Sharing Stories of Women in Wind." The stories range from a biography of Oya, the West African goddess of the wind, to former winners of the Rudd Mayer fellowship to officers of the organization, to other women who have made important contributions to the wind industry.

An important feature of the WoWE Web site is a section on careers in wind energy for women. The section provides information wind careers in general, educational opportunities, an introduction to careers, suggestions for mentoring, webinars on topics of importance in the field, and opportunities to apply for awards and fellowships.

In additional to individual memberships, WoWE makes corporate memberships available. Some of the organization's current corporate sponsors are the American Wind Energy Association, North American Windpower, *enerG* magazine, GE Energy, and DNV GL offshore wind systems.

World Wind Energy Association

Charles-de-Gaulle-Str. 5
53113 Bonn
Germany
Phone: +49 228 369 40 80
Fax: +49 228 369 40 84
Email: http://wwindea.org/home/index.php?option=com_performs&formid=4&Itemid=71
URL: http://wwindea.org/home/index.php

The World Wind Energy Association (WWEA) was formed on July 1, 2001, in Copenhagen, Denmark, by a coalition of national wind agencies and related groups. Founding members were the Australian Wind Energy Association, Brazilian Wind Energy Association, Eurosolar Denmark, Egyptian Wind Energy Association, German Wind Energy Association, Indian Wind Turbine Manufacturers' Association, Japan Wind Energy

Association, Wind Energy Research Group Korea, Norwegian Wind Energy Association, Russian Academy of Sciences Kola Science Center, South African Wind Energy Association, and Carl-Duisberg-Gesellschaft e.V.

WWEA currently has three categories of membership: ordinary members, scientific members, and corporate members. The association also has a category known as observers who are not members but who participate in other ways in WWEA activities. Examples of the more than eighty ordinary members are the African Wind Energy Association (AfriWEA), Societe Algerienne de l'Energie Solaire (Algeria), Somaliland Energy for Sustainable Development Organization (Somalia), International Agri Cooperative & Development Organisation (Sri Lanka), and the Ukrainian Wind Energy Association (Ukraine). Some current scientific members are the Institute for Environment and Development Studies of Bangladesh, National Windpower Technology Research Center of China, Institute for Energy Technology of the University of Malta, and Technical University of Kosice, Center of Alternative Energy Resources (Slovakia). Examples of corporate members are Green Frog Energy (Australia), Simex Technologies Canada, Shanghai Hi-tech Control System (China), and Global Wind Systems Inc. (United States).

WWEA was founded on five major principles:

1. Wind energy shall serve as one cornerstone and driving force for the immediate implementation of the world energy system driven by renewable energies to completely substitute fossil and nuclear sources.
2. Global dissemination of grid connected and stand-alone wind energy solutions should rely on experience gained from the most successful implementation strategies, based on favourable legal, political and social framework conditions as initiated by national associations. Local and rural communities and people should be involved and should benefit directly.

3. WWEA shall stimulate and back the foundation of national and regional wind energy associations and encourage national governments to set ambitious targets and political frameworks for priority strategies in favour of a fast and sustainable development of all renewable energies.
4. WWEA will organise World Wind Energy Conferences and further international events for mobilising a wide range of the different wind energy applications.
5. WWEA will play an active role in the World Council for Renewable Energies in order to work for a full substitution of all polluting and hazardous waste causing energies. (From the WWEA press release of June 30, 2001; http://www.wwindea.org/home/index.php?option=com_content&task=view&id=148&Itemid=40.)

WWEA is governed by a board consisting of the association president and ten vice presidents from each of the continents. Day-to-day operations are run out of the association's headquarters office in Bonn by a secretary general, who is currently Stefan Gsänger.

Among the most important activities sponsored by the WWEA are its conferences and other meetings. The highlight each year is the annual WWEA conference, the first of which was held in Berlin in 2002. Since then, other conferences have been held in Cape Town, Beijing, Melbourne, New Delhi, Mar del Plata (Argentina), Kingston (Ontario), Jeju Island (South Korea), Istanbul, and Havana. WWEA also sponsors and cosponsors other international meeting on wind power and related topics, such as the World Summit for Small Wind, Wind Power Asia, Clean Energy Expo, World Bioenergy, Local Renewables, Renewable Energy Storage, and Advanced Wind Technology and Investment.

Two areas of special interest to WWEA are small wind issues and wind energy technology. The organization provides information and guidance on both topics, as well as organizing conference and other meetings on the subjects. In the former area,

WWEA provides a developer's guide that includes information and guidance on dealer and distributor services, standards, consumer labeling, certification and testing, policies, and a database of potential partners. It also offers a buyer's guide, a business directory, and a forum on small wind issues.

In the area of wind energy technology, WWEA provides an excellent general and technical introduction to the topic of wind energy, with sections on the history of wind power, planning for facilities, operation and maintenance, grid integration and storage, and special applications.

5 Data and Documents

Introduction

One can gain an insight into the history and nature of a nation's attitudes toward wind energy by examining its documentary history, laws, acts, regulations, reports, court cases, and other legal documents dealing with the topic. Other information can be gained from a review of statistical trends relating to the production and consumption of wind energy. The "Data" section of this chapter contains a selection of such statistical data. The second part of this chapter contains a selection of excerpts from some of those documents in the history of the United States. It should be noted in these documents that wind energy is essentially never treated separately from other forms of alternative energy but is included in legislation more broadly aimed at that topic.

Data

Table 5.1. Wind Energy Generation (megawatt hours), United States, 2011 and 2012. The Energy Information Administration (EIA) of the U.S. Department of Energy collects, analyzes, and publishes data on a regular basis about all aspects of the production and consumption of all forms of energy in the United States. This table summarizes data on the availability of electrical energy produced by wind power in 2011 and 2012, the latest years for which such data are available.

An offshore wind farm near Copenhagen, Denmark. (Rodik/Dreamstime.com)

Table 5.1 Wind Energy Generation (in megawatt hours), United States, 2011 and 2012

Region	2011	2012	Percentage Change
New England	870	1,294	+48.6
Middle Atlantic	4,633	5,132	+10.8
East North Central	11,341	14,612	+28.8
West North Central	31,288	37,561	+20.0
South Atlantic	1,378	1,611	+16.9
East South Central	53	47	–10.6
West South Central	36,153	40,372	+11.7
Mountain	15,317	17,080	+11.5
Pacific Contiguous	18,790	22,697	+20.8
Pacific Noncontiguous	353	416	+17.7
U.S. Total	120,177	140,822	+17.2

Source: Table 3.17. Net Generation from Wind. Annual Energy Outlook 2014. http://www.eia.gov/electricity/annual/html/epa_03_17.html. Accessed on March 5, 2014.

Table 5.2 States with Largest Installed Wind Power Capacity (in megawatt hours), 2011–2012. EIA also collects data on wind-generated electrical energy for each of the 50 states, data summarized in this table for 2011 and 2012 also.

State	2011	2012	Percentage Change
Texas	30,548	+32,214	+5.5
Iowa	10,709	14,032	+31.0
California	7,752	9,754	+25.8
Oklahoma	5,605	8,158	+45.5
Illinois	6,213	7,727	+24.4
Minnesota	6,726	7,615	+13.2
Washington	6,262	6,600	+5.4
Oregon	4,775	6,343	+32.8
Colorado	5,200	5,969	+14.8
North Dakota	5,236	5,275	+0.7
Kansas	3,720	5,195	+39.7
Wyoming	4,612	4,369	–5.3

Source: Table 3.17. Net Generation from Wind. Annual Energy Outlook 2014. http://www.eia.gov/electricity/annual/html/epa_03_17.html. Accessed on March 5, 2014.

Table 5.3 Energy Generation by Wind in the United States and Worldwide in (billion kilowatt-hours), 1983–2012. Wind power-generating capacity has increased dramatically since the early 1980s, when the first wind farms were being established in the United States and around the world. This table shows some of the data for this change.

Year	United States	Worldwide
1983	0.003	0.030
1984	0.006	0.050
1985	0.006	0.073
1986	0.004	0.145
1987	0.004	0.206
1988	0.001	0.343
1989	2.112	2.598
1990	2.789	3.536
1991	2.951	4.097
1992	2.888	4.611
1993	3.006	5.558
1994	3.447	7.284
1995	3.164	7.936
1996	3.234	9.289
1997	3.288	12.135
1998	3.026	16/109
1999	4.488	21.242
2000	5.593	31.364
2001	6.737	38.367
2002	10.354	52.766
2003	11.187	64.306
2004	14.144	84.136
2005	17.811	104.021
2006	26.589	131.830
2007	34.450	170.563
2008	55.363	220.298
2009	73.886	276.045
2010	94.652	341.532
2011	120.177	446.334
2012	140.089	n/a

Source: International Energy Statistics. U.S. Energy Information Administration. http://www.eia.gov/cfapps/ipdbproject/iedindex3.cfm?tid=2&pid=37&aid=12&cid=&syid=1980&eyid=2012&unit=BKWH. Accessed on March 9, 2014.

Table 5.4 Wind Energy Budget Allocations, FY 2012, and Requests, FY 2014 (in thousands of dollars). Federal financing for wind projects is spread out through a number of federal agencies. By far the largest portion of that budget comes from the Department of Energy, whose budget allocations for two recent years are shown in this table.

Item	FY 2012 Allocation	FY 2014 Request
Technology Development and Testing	73,054	99,000
Offshore Wind	34,442	46,000
Technology Components R&D	19,525	6,500
Plant Optimization	0	23,500
Manufacturing	1,389	6,000
Testing Infrastructure	15,811	5,000
Distributed Wind Technology	1,887	7,500
Wind Technology Incubator	0	4,500
Technology Application	18,759	36,000
Resource Characterization	6,816	12,500
Grid Optimization	4,916	10,500
Eliminate Market Barriers	7,027	13,000
NREL User Facility	0	9,000
Total, Wind Energy	91,813	144,000

*FY 2013 total wind allocation: $93,825,000

Source: Wind Energy—FY 2014 Budget Request. Office of Energy Efficiency and Renewable Energy. http://www1.eere.energy.gov/office_eere/pdfs/electricity_stakeholder_pres_0513.pdf. Accessed on March 9, 2014. For a complete description of the categories shown in the table, see this source.

Table 5.5 Candidate Wind Turbine Sites. In 1976, the U.S. Department of Energy began collecting data on possible sites for the installation of wind turbines in the United States. That study was continued in 1980, after which a list of so-called candidate sites has been maintained by the department. A selection of those sites with relevant characteristics is shown in this table.

Location	State	Elevation (m)	Anemometer Height (m)	Average Annual Wind Speed (m/s)	Average Annual Wind Power Density (W/m2)
Cold Bay	AK	29	21.8	7.3	496
Point Arena	CA	21	45.7	6.5	322
San Gorgonio Pass	CA	344	45.7	7.7	712
Ilio Point, Molokai	HI	61	45.7	10.9	1,032
Kahua Ranch, Oahu	HI	1,030	45.7	11.3	1,528
Nantucket Island	MA	12	45.7	9.1	697
Livingston	MN	1,420	45.7	8.4	794
Wells	NV	2,268	45.7	7.8	408
Montauk Point	NY	2	45.7	7.2	436
Finley	ND	472	45.7	9.1	737
Block Island	RI	14	45.7	7.4	407
Amarillo	TX	1,091	45.7	8.1	464
Diablo Dam	WA	500	45.7	5.1	159
Bridger Butte	WY	2,290	45.7	8.4	589

Source: Table E 1. U.S. Department of Energy Candidate Wind Turbine Sites. http://rredc.nrel.gov/wind/pubs/atlas/tables/E 1T.html#lbsign. Accessed on April 16, 2014.

Table 5.6 **Energy Consumption from Renewables in the United States, Actual and Projected. (quadrillion Btu) This table shows the amount of energy produced from renewable sources in the period from 2009 to 2013, with projections for 2014 and 2015.**

Sector	2009	2010	2011	2012	2013	2014	2015
Hydroelectric	2.650	2.521	3.085	2.606	2.529	2.622	2.596
Wood Biomass	0.180	0.196	0.182	0.190	0.207	0.258	0.273
Waste Biomass	0.261	0.264	0.255	0.262	0.258	0.267	0.274
Wind	**0.721**	**0.923**	**1.167**	**1.339**	**1.595**	**1.663**	**1.789**
Geothermal	0.146	0.148	0.149	0.148	0.157	0.160	0.163
Solar	0.009	0.012	0.017	0.040	0.085	0.149	0.175
Total	3.967	4.064	4.855	4.586	4.831	5.078	5.270

Source: "U.S. Renewable Energy Consumption (Quadrillion Btu)." Short term Energy and Summer Fuels Outlook. U.S. Energy Information Administration. http://www.eia.gov/forecasts/steo/tables/?tableNumber=24#startcode=2009. Accessed on April 17, 2014.

Table 5.7 Global Offshore Wind Capacity, as of 2013. The U.S. Department of Energy conducts an annual survey on the status of offshore wind facilities in the United States and the rest of the world. This table summarizes the general data for offshore wind farms globally.

Country	Number of Projects	Total Capacity (MW)	Total Number of Turbines
Asia			
China	13	365	138
Japan	5	28	16
South Korea	2	5	2
Europe			
Belgium	3	380	91
Denmark	16	875	406
Finland	3	32	11
Germany	7	286	69
Ireland	1	25	7
Netherlands	4	247	128
Norway	1	2	1
Portugal	1	2	1
Sweden	5	164	75
United Kingdom	24	2,874	850
Total	85	5,284	1,795

Source: "Summary of Installed Global Offshore Capacity through 2012." Offshore Wind Market and Economic Analysis. http://www1.eere.energy.gov/wind/pdfs/offshore_wind_market_and_economic_analysis_10_2013.pdf. Accessed on April 18, 2014.

Table 5.8 Offshore Wind Policy Highlights for 2013. The development of new wind facilities, both in the United States and the rest of the world, depends to a considerable degree on efforts by national and state governments to promote such developments. The 2013 Offshore Wind Market and Economics Analysis, sponsored by the U.S. Department of Energy, explored the types of incentives state and national governments have been using to promote offshore wind developments. This table summarizes some of those efforts.

» **Policies that address cost-competitiveness**

The U.S. PTC and ITC were extended for projects that begin construction by year-end 2013. The 50 percent first-year bonus depreciation allowance was also extended for one year.

(*Continued*)

Table 5.8 (*Continued*)

The U.S. DOE announced seven projects that will receive up to $4 million each to complete engineering and planning as the first phase of the Offshore Wind Advanced Technology Demonstration Program.

The Maryland Offshore Wind Energy Act of 2013 established Offshore Wind Renewable Energy Credits (ORECs) for up to 200 MW.

The Maine legislature passed a bill that reopened the bidding process for ratepayer subsidies to offshore wind projects.

The United Kingdom announced the strike prices for land-based and offshore wind generation through 2019, which should expedite the development of Round 3 projects.

Spain has made various reductions to its FiT with, in some cases, retroactive effects on existing projects.

» **Policies that address infrastructure challenges**

The New German Energy Act clarifies the compensation that projects impacted by grid delays are entitled to; that law is expected to resolve the grid construction delays.

» **Policies that address regulatory challenges**

BOEM held the first two competitive lease sales for renewable energy in U.S. federal waters off the coasts of Rhode Island and Virginia.

Illinois passed the Lake Michigan Wind Energy Act, which requires the Illinois DNR to develop a detailed offshore wind energy siting matrix for Lake Michigan.

Denmark initiated feasibility studies for six areas that have been identified for offshore [wind facilities].

Source: "Offshore Wind Market and Economic Analysis." U.S. Department of Energy, 48. http://www1.eere.energy.gov/wind/pdfs/offshore_wind_market_and_economic_analysis_10_2013.pdf. Accessed on April 18, 2014.

Explanation of acronyms:

PTC: Renewable Electricity Production Tax Credit
ITC: Business Energy Investment Tax Credit
DOE: Department of Energy
FiT: Feed-in tariffs
BOEM: Bureau of Ocean Energy Management
DNR: Department of Natural Resources

Table 5.9 Postsecondary Education in Wind Energy. Nearly 200 degree, certificate, and other educational programs are now available in wind energy in a variety of postsecondary educational institutions in the United States. This table provides basic information on a selection of those institutions. See the source reference at the end of the table for a complete list of such institutions.

State	Institution	Degree/Certificate
CA	Rio Hondo Community College	AAS Degree in Alternative Energy Technology
CA	Cerro Coso Community College	Industrial Technology Wind and Solar pathways
CA	San Diego State University	Online Certificate in Green Energy Management
CO	Redstone College	AA Degree in Wind Energy Technology
CO	Ecotech Institute	Wind Energy Technology
IA	University of Iowa	Wind Power Management
IA	Des Moines Area Community College	Wind Turbine Technician
ID	College of Southern Idaho	Wind Energy Technician
IL	Danville Area Community College	Wind Turbine Technician
MA	University of Massachusetts at Amherst	Certificates and Degrees in a Variety of Programs
MD	Hagerstown Community College	AAS Degree in Alternative Energy Technology
ME	Northern Maine Community College	AAS Degree in Wind Power Technology
MI	Grand Rapids Community College	Competent Climber and Tower Rescue
MN	Minnesota West Community & Technical College	Wind Energy Mechanic Wind Energy Technology
MO	Pinnacle Career Institute	Wind Turbine Technician
NC	North Carolina State University	Diploma in Wind Energy
ND	Lake Region State College	Wind Energy Technician
NM	Clovis Community College	AAS Degree in Industrial Technology with a Concentration in Wind Energy

(Continued)

Table 5.9 (*Continued*)

State	Institution	Degree/Certificate
NV	Truckee Meadows Community College	AAS Degree in Renewable Energy
NY	Clinton Community College	Wind Energy and Turbine Technology
NY	State University of New York at Canton	BS Degree in Alternative and Renewable Energy Systems
OH	Lorain County Community College	Wind Turbine Technology
OK	Oklahoma State University	AAS Degree in Wind Turbine Technology
OR	Columbia Gorge Community College	Renewable Energy Technology
PA	Saint Francis University	Renewable Energy Business Development Certificate
SD	Mitchell Technical Institute	Wind Turbine Technology
TX	Texas Tech University	Wind Science and Engineering Certificate in Energy Law
WI	Lakeshore Technical College	Wind Energy Technology
WV	Eastern West Virginia Community & Technical College	AAS Degree in Wind Energy Turbine Technology
WY	Laramie County Community College	Wind Energy

Source: Compiled from Wind Energy Education and Training Programs. U.S. Department of Energy. Energy Efficiency and Renewable Energy. http://www.windpoweringamerica.gov/schools/education/education_training.asp. Accessed on April 18, 2014.

Documents

Energy Policy and Conservation Act of 1975

U.S. policy for the use of wind energy had a very modest beginning in 1975 with the adoption of the Energy Policy and Conservation Act of 1975. That act was inspired largely by the sudden risk in oil prices in the early 1970s but, for the first time, acknowledged the possible value of using alternative sources of energy in the United States. Although wind power is not mentioned specifically in the act, the following section shows that the U.S. Congress had begun to think in terms of the role of alternative energy sources at this point in history. The very limited statement relevant to the history of wind energy use is indicated in bold in the following selection.

Individual Energy Efficiency Improvement Targets

Sec. 374.

(a) Within one year after the date of enactment of this Act, the Administrator shall set an industrial energy efficiency improvement target for each of the 10 most energy-consumtive *[sic]* industries identified under section 373. Each such target—

  (1) shall be based upon the best available information
  (2) shall be established at the level which represents the maximum feasible improvement in energy efficiency which such industry can achieve by January 1, 1980, and
  (3) shall be published in the Federal Register, together with a statement of the basis and justification for each such target.

(b) In determining maximum feasible improvement under subsection (a) and under subsection (c), the Administrator shall consider—

  (1) the objectives of the program established under section 372,

(2) the technological feasibility and economic practicability of utilizing alternative operating procedures and more energy efficient technologies. . . .

Source: Energy Policy and Conservation Act. U.S. Code 42 (1975): §§ 6201 et seq.

Renewable Energy and Energy Efficiency Competitiveness Act 1989

The first piece of federal legislation in the United States to speak very specifically about the goals and objectives of programs for the promotion of alternative energy sources was the Renewable Energy and Energy Efficiency Competitiveness Act of 1989. That act expressed these goals primarily for wind, photovoltaic, and solar thermal energy programs. The section relating to wind energy is as follows:

Sec. 4. National Goals for Federal Wind, Photovoltaic, and Solar Thermal Programs.

(a) NATIONAL GOALS- The following are declared to be the national goals for the wind, photovoltaic and solar thermal energy programs being carried out by the Secretary.

(1) WIND—(A) In general, the goals for the Wind Energy Research Program include improving design methodologies and developing more reliable and efficient wind turbines to increase the cost competitiveness of wind energy. Research efforts shall include—

(i) activities that address near-term technical problems and assist private-sector exploitation of market opportunities of the wind energy industry;

(ii) developing technologies such as advanced airfoils and variable speed generators to increase wind turbine output and reduce maintenance costs by decreasing structural stress and fatigue;

(iii) increasing the basic knowledge of aerodynamics, structural dynamics, fatigue and electrical systems interactions as applied to wind energy technology; and

(iv) improving the compatibility of electricity produced from wind farms with conventional utility needs.

(B) Specific goals for the Wind Energy Research Program shall be to—

(i) reduce average wind energy costs to 3 to 5 cents per kilowatt hour by 1995;

(ii) reduce capital costs of new wind energy systems to $500 to $750 per kilowatt of installed capacity by 1995;

(iii) reduce operation and maintenance costs for wind energy systems to less than 1 cent per kilowatt hour by 2000; and

(iv) increase capacity factors for new wind energy systems to 25 to 30 per centum by 1995.

Source: Public Law 101–218—December 11, 1989. Government Printing Office. Accessed March 5, 2014. (Formatting of the bill and act vary from publication to publication.)

Interim Guidelines to Avoid and Minimize Wildlife Impacts from Wind Turbines (2003)

Although the federal government has adopted relatively few laws and regulations dealing specifically with the construction and use of wind power systems, a number of federal agencies have developed rules, regulations, guidelines, and suggestions about this topic. An example is the set of guidelines developed by the U.S. Fish and Wildlife Service to reduce the impact of wind turbines on wildlife. An excerpt from this document follows:

Site Development Recommendations

The following recommendations apply to locating turbines and associated structures within WRAs *[wind resource areas]* selected for development of wind energy facilities:

1. Avoid placing turbines in documented locations of any species of wildlife, fish, or plant protected under the Federal Endangered Species Act.
2. Avoid locating turbines in known local bird migration pathways or in areas where birds are highly concentrated, unless mortality risk is low (e.g., birds present rarely enter the rotor-swept area). Examples of high concentration areas for birds are wetlands, State or Federal refuges, private duck clubs, staging areas, rookeries, leks, roosts, riparian areas along streams, and landfills. Avoid known daily movement flyways (e.g., between roosting and feeding areas) and areas with a high incidence of fog, mist, low cloud ceilings, and low visibility.
3. Avoid placing turbines near known bat hibernation, breeding, and maternity/nursery colonies, in migration corridors, or in flight paths between colonies and feeding areas.
4. Configure turbine locations to avoid areas or features of the landscape known to attract raptors (hawks, falcons, eagles, owls). For example, Golden Eagles, hawks, and falcons use cliff/rim edges extensively; setbacks from these edges may reduce mortality. Other examples include not locating turbines in a dip or pass in a ridge, or in or near prairie dog colonies.
5. Configure turbine arrays to avoid potential avian mortality where feasible. For example, group turbines rather than spreading them widely, and orient rows of turbines parallel to known bird movements, thereby decreasing the potential for bird strikes. Implement appropriate storm water management *practices that do not create attractions for* birds, and maintain contiguous habitat for area-sensitive species (e.g., Sage Grouse).

6. Avoid fragmenting large, contiguous tracts of wildlife habitat. Where practical, place turbines on lands already altered or cultivated, and away from areas of intact and healthy native habitats. If not practical, select fragmented or degraded habitats over relatively intact areas.
7. Avoid placing turbines in habitat known to be occupied by prairie grouse or other species that exhibit extreme avoidance of vertical features and/or structural habitat fragmentation. In known prairie grouse habitat, avoid placing turbines within 5 miles of known leks (communal pair formation grounds).
8. Minimize roads, fences, and other infrastructure. All infrastructure should be capable of withstanding periodic burning of vegetation, as natural fires or controlled burns are necessary for maintaining most prairie habitats.
9. Develop a habitat restoration plan for the proposed site that avoids or minimizes negative impacts on vulnerable wildlife while maintaining or enhancing habitat values for other species. For example, avoid attracting high densities of prey animals (rodents, rabbits, etc.) used by raptors.
10. Reduce availability of carrion by practicing responsible animal husbandry (removing carcasses, fencing out cattle, etc.) to avoid attracting Golden Eagles and other raptors.

Turbine Design and Operation Recommendations

1. Use tubular supports with pointed tops rather than lattice supports to minimize bird perching and nesting opportunities. Avoid placing external ladders and platforms on tubular towers to minimize perching and nesting. Avoid use of guy wires for turbine or meteorological tower supports. All existing guy wires should be marked with recommended bird deterrent devices (Avian Power Line Interaction Committee 1994).

2. If taller turbines (top of the rotor-swept area is >199 feet above ground level) require lights for aviation safety, the minimum amount of pilot warning and obstruction avoidance lighting specified by the Federal Aviation Administration (FAA) should be used (FAA 2000). Unless otherwise requested by the FAA, only white strobe lights should be used at night, and these should be the minimum number, minimum intensity, and minimum number of flashes per minute (longest duration between flashes) allowable by the FAA. Solid red or pulsating red incandescent lights should not be used, as they appear to attract night-migrating birds at a much higher rate than white strobe lights.
3. Where the height of the rotor-swept area produces a high risk for wildlife, adjust tower height where feasible to reduce the risk of strikes.
4. Where feasible, place electric power lines underground or on the surface as insulated, shielded wire to avoid electrocution of birds. Use recommendations of the Avian Power Line Interaction Committee (1994, 1996) for any required above-ground lines, transformers, or conductors.
5. High seasonal concentrations of birds may cause problems in some areas. If, however, power generation is critical in these areas, an average of three years monitoring data (e.g., acoustic, radar, infrared, or observational) should be collected and used to determine peak use dates for specific sites. Where feasible, turbines should be shut down during periods when birds are highly concentrated at those sites.
6. When upgrading or retrofitting turbines, follow the above guidelines as closely as possible. If studies indicate high mortality at specific older turbines, retrofitting or relocating is highly recommended.

Source: Service Interim Guidelines on Avoiding and Minimizing Wildlife Impacts from Wind Turbines. Fish and Wildlife

Service. http://www.fws.gov/habitatconservation/wind.pdf. Accessed on March 7, 2014.

Federal Wind Program: Goals (2006)

The U.S. Department of Energy (DOE) Wind Program was established in 1975 under the auspices of the department's predecessor, the Energy Research and Development Administration (ERDA). Ever since that time, the Wind Program has been the lead federal agency for the promotion and development of wind facilities in the United States. A major feature of the Wind Program has been the 20% Wind Energy by 2030 program, which calls for the production of a fifth of the nation's electricity needs by wind power by the year 2030. The specific goals of that program are as follows:

- By 2007, reduce the cost of electricity from distributed wind systems to 10-15 cents/kWh in Class 3 wind resources.
- By 2010, facilitate the installation of at least 100 MW of wind energy in 30 states.
- By 2012, reduce the cost of electricity from large wind systems in Class 4 winds to 3.6 cents/kWh for onshore systems.
- By 2012, complete program activities addressing electric power market rules, interconnection impacts, operating strategies, and system planning needed for wind energy to compete without disadvantage to serve the Nation's energy needs.
- By 2014, reduce the cost of electricity from large wind systems in Class 6 winds to 5 cents/kWh for shallow water (depths up to 30 meters) offshore systems (from a baseline of 9.5 cents/kWh in 2005).
- By 2016, reduce the cost of electricity from large wind systems in Class 6 winds to 5 cents/kWh for transitional (depths up to 60 meters) offshore systems.

Note: Wind systems are divided into five "classes," based on their wind power density and operating wind speed. See the report "20% Wind Energy by 2030," Table B-7.

Source: "Wind Power Today." U.S. Department of Energy. Energy Efficiency and Renewable Energy, 2. http://www.nrel.gov/docs/fy06osti/39479.pdf. Accessed on April 17, 2014.

Initiative 937, Washington State (2006)

Concerns about the depletion of fossil fuel supplies in the United States have led a number of states to adopt legislation requiring energy utilities to produce some fraction of their product by renewable sources by some set future deadline. As of late 2014, twenty states and the District of Columbia have adopted rules, regulations, and/or legislation of this kind. An example of the type of regulation that has been adopted is Initiative 937, adopted by the voters of the state of Washington in November 2006. The initiative requires power facilities to produce at least 15 percent of the energy they generate by renewable sources, such as wind energy. A portion of that initiative read as follows:

AN ACT Relating to requirements for new energy resources; adding a new chapter to Title 19 RCW [Revised Code of Washington]; and prescribing penalties.

Be it Enacted by the People of the State of Washington:

NEW SECTION. Sec. 1. INTENT. This chapter concerns requirements for new energy resources. This chapter requires large utilities to obtain fifteen percent of their electricity from new renewable resources such as solar and wind by 2020 and undertake cost-effective energy conservation.

NEW SECTION. Sec. 2. DECLARATION OF POLICY. Increasing energy conservation and the use of appropriately sited renewable energy facilities builds on the strong foundation of low-cost renewable hydroelectric generation in Washington state and will promote energy independence in the state and the Pacific Northwest region. Making the most of our plentiful local resources will stabilize electricity prices for Washington residents, provide economic benefits for Washington counties and farmers, create high-quality jobs in Washington, provide opportunities for training apprentice workers in the renewable energy field, protect clean air and water, and position Washington state as a national leader in clean energy technologies.

* * *

(2)(a) Each qualifying utility shall use eligible renewable resources or acquire equivalent renewable energy credits, or a combination of both, to meet the following annual targets:

- (i) At least three percent of its load by January 1, 2012, and each year thereafter through December 31, 2015;
- (ii) At least nine percent of its load by January 1, 2016, and each year thereafter through December 31, 2019; and
- (iii) At least fifteen percent of its load by January 1, 2020, and each year thereafter.

Source: Initiative 937. Secretary of State, State of Washington, http://www.secstate.wa.gov/elections/initiatives/text/i937.pdf. Accessed on March 8, 2014. For more detailed information on state laws of this type, see "Renewable Energy Standards at Work in the States." Union of Concerned Scientists. http://www.ucsusa.org/assets/documents/clean_energy/RES_in_the_States_Update.pdf. Accessed on March 8, 2014.

Rankin v. FPL Energy, LLC, 266 S.W.3d 506 (2008)

Just as large-scale wind power for the production of energy is a relatively new technology, so is case law dealing with a number of potential issues that may arise as a result of the expansion of wind farms and similar facilities. As of late 2014, one of the most influential cases dealing with the effect of wind farms on nearby residents is Rankin v. FPL, LLC, *summarized here. In the case, a group of individuals and one corporation in east Texas filed suit against the Horse Hollow Wind Farm, built and operated by the FPL Energy Company. Plaintiffs claimed that the farm was a "public nuisance" because its presence imposed on their own privacy. They argued specifically that they objected to the visual appearance of the wind turbines. The trial court found in favor of the defendants, and the plaintiffs proceeded to appeal that decision to the Texas Eleventh Court of Appeals. That court upheld the trial court's decision (with the exception of determining the parties' responsibility for trial costs), with the following reasoning. Ellipses (. . .) indicate the omission of references, citations, or other extraneous material.*

In practice, successful nuisance actions typically involve an invasion of a plaintiff's property by light, sound, odor, or foreign substance. For example, in Pascouet, floodlights that illuminated the plaintiffs' backyard all night and noisy air conditioners that interfered with normal conversation in the backyard, that could be heard indoors, and that interrupted plaintiffs' sleep constituted a nuisance. . . .

Texas courts have not found a nuisance merely because of aesthetical-based complaints. In Shamburger v. Scheurrer . . . , the defendant began construction of a lumberyard in a residential neighborhood. Neighboring homeowners filed suit and contended that the lumberyard would be unsightly, unseemly, and have ugly buildings and structures. The court held that this did not constitute a nuisance. . . .

Plaintiffs' summary judgment evidence makes clear that, if the wind farm is a nuisance, it is because Plaintiffs' emotional response to the loss of their view due to the presence of numerous wind turbines substantially interferes with the use and enjoyment of their property. The question, then, is whether Plaintiffs' emotional response is sufficient to establish a cause of action. One Texas court has held that an emotional response to a defendant's lawful activity is insufficient. . . .

If Plaintiffs have the right to bring a nuisance action because a neighbor's lawful activity substantially interferes with their view, they have, in effect, the right to zone the surrounding property. Conversely, we realize that Plaintiffs produced evidence that the wind farm will harm neighboring property values and that it has restricted the uses they can make of their property. FPL's development, therefore, could be characterized as a condemnation without the obligation to pay damages.

Texas case law has balanced these conflicting interests by limiting a nuisance action when the challenged activity is lawful to instances in which the activity results in some invasion of the plaintiff's property and by not allowing recovery for emotional reaction alone. Altering this balance by recognizing a new cause of action for aesthetical impact causing an emotional injury is beyond the purview of an intermediate appellate court. . . .

We reverse the judgment of the trial court in part and remand the cause for determination of the allocation of taxable costs. We affirm the judgment of the trial court in all other respects.

Source: *Rankin v. FPL Energy, LLC,* 266 S.W.3d 506.

Wildlife Guidelines for Wind Power Facilities, State of Arizona (2009)

One of the ongoing concerns about the construction and use of wind power facilities is their potential harm to wildlife populations in the siting area. A relatively small number of states have taken action to deal with this issue by adopting voluntary or mandatory guidelines for the siting of wind farms in such a way as to protect bats, birds, and other forms of wildlife within their borders. In most cases, these regulations are detailed and comprehensive documents running dozens of pages of rules and regulations. An example of the type of guidelines provided can be found in a document issued in 2009 by the Arizona Game and Fish Department, reproduced here. According to this document, some of the information to be considered in assessing the impact of wind power facilities on wildlife are:

1. Which species of bats and birds use the project area and how do their numbers vary throughout the year?
2. Are any of the following known, or likely to occur, on or near the proposed project site?

   (Note: "Near" refers to a distance within the area used by an animal in the course of its normal movements and activities.)

   a. Species listed as federal "Threatened" or "Endangered" (or candidates for such listing)?
   b. Special status bat or bird species?
   c. Bald or Golden eagles?
3. Is the site near a raptor nest, or are large numbers of raptors known or likely to occur at or near the site during portions of the year?
4. Is the site on or near important staging or wintering bird areas?
5. Are there prey species such as prairie dog colonies and high insect prey bases attracting wildlife populations to the area?

   (Note: Species that may not appear to have a direct conflict with wind development may result in greater impacts to raptors due to the area's importance as a foraging site.)
6. Is the site likely to be used by birds whose behaviors include flight displays (e.g. common nighthawks, horned larks) or by species whose foraging tactics put them at risk of collision (e.g. contour hunting by golden eagles)?
7. Is the site near a known or potential bat roost, recognizing some species of bats will fly over 20 miles each way to forage?

8. Are there physical features such as ridgelines, cliff faces, caves (or cracks and fissures), unique vegetation communities, riparian areas, water or forage sources attracting and concentrating wildlife populations (e.g. foraging, roosting, breeding, or cover habitat)? Is the site near a known or likely migrant stopover site?
9. Is the site regularly characterized by seasonal weather conditions such as dense fog or low cloud cover which may increase collision risks to bats and birds, and do these events occur at times when birds may be concentrated?
10. If the site has characteristics which concentrate wildlife, what potential design and mitigation measures could reduce impacts?

Source: Guidelines for Reducing Impacts to Wildlife from Wind Energy Development in Arizona, Arizona Game and Fish Department, Revised November 23, 2009. http://www.azgfd.gov/hgis/pdfs/WindEnergyGuidelines.pdf. Accessed on March 8, 2014.

American Recovery and Reinvestment Act of 2009 (P.L. 111-5)

The most recent federal legislation dealing with wind energy is the American Recovery and Reinvestment Act of 2009. Like previous bills, this act lumps a number of forms of renewable energy together in its provisions. The specific provisions in support of wind energy are essentially technical in nature, with tax credits and other financial incentives provided for the development and implementation of wind facilities. A key provision of the act is the following:

Part III—Energy Conservation Incentives

Sec. 1121. Extension and Modification of Credit for Nonbusiness Energy Property.

(a) In General.—Section 25C is amended by striking subsections (a) and (b) and inserting the following new subsections:

"(a) Allowance of Credit.—In the case of an individual, there shall be allowed as a credit against the tax imposed by this chapter for the taxable year an amount equal to 30 percent of the sum of—

"(1) the amount paid or incurred by the taxpayer during such taxable year for qualified energy efficiency improvements, and

"(2) the amount of the residential energy property expenditures paid or incurred by the taxpayer during such taxable year.

"(b) Limitation.—The aggregate amount of the credits allowed under this section for taxable years beginning in 2009 and 2010 with respect to any taxpayer shall not exceed $1,500."

Source: Public Law 111–5—February 17, 2009. Government Printing Office. http://www.gpo.gov/fdsys/pkg/PLAW-111publ5/pdf/PLAW-111publ5.pdf. Accessed on March 5, 2014. For more information on court cases related to the impact of wind power facilities, see Barnard 2014, in Chapter 6 of this book.

Sowers v. Forest Hills Subdivision, 294 P. 3d 427 (2013)

Legal cases have arisen not only about the siting of large wind farms but also about the construction of small wind turbines for the generation of electricity for individual residences. Perhaps the most significant case of this type is one brought by the residences of a community in Las Vegas, Nevada. The case eventually reached the Nevada Supreme Court, which became the highest court in the nation to rule on such an issue. The court's decision read in part as follows:

In this appeal, we address whether the district court properly concluded that, under the particular circumstances and surroundings of the case, a proposed residential wind turbine would constitute a nuisance warranting a permanent injunction against its construction. Below, respondents Forest Hills Subdivision, Ann Hall, and Karl Hall collectively, the Halls) sought to permanently enjoin their neighbor, appellant Rick Sowers, from constructing a wind turbine on his residential property, asserting that the proposed turbine would constitute a nuisance. The district court agreed and granted the permanent injunction.

We conclude that, in this case, substantial evidence exists to support the district court's conclusion that the proposed wind turbine constitutes a nuisance. We also determine that the wind turbine at issue would create a nuisance in fact. In reaching our conclusion, we hold that the aesthetics of a wind turbine alone are not grounds for finding a nuisance. However, we conclude that a nuisance in fact may be found when the aesthetics are combined with other factors, such as noise, shadow flicker, and diminution in property value. In this case, the district court heard testimony about the aesthetics of the proposed wind turbine, the noise and shadow flicker it would create, and its potential to diminish surrounding property values. Based on this evidence, we conclude that substantial evidence supports the district court's finding that the proposed residential wind turbine would be a nuisance in fact. Thus, we affirm the order granting a permanent injunction prohibiting its construction.

Source: *Sowers v. Forest Hills Subdivision.* Nevada Supreme Court. http://statecasefiles.justia.com/documents/nevada/supreme-court/129-nev-adv-op-no-9.pdf?ts=1370901315. Accessed on April 17, 2014.

Alliance to Protect Prince Edward County v. Director, Ministry of the Environment (2013)

Efforts to block the construction and/or operation of wind power facilities have been made over at least the past decade in courts and other judicial bodies at all levels in many countries of the world. According to one authority in the field, the United States is (perhaps) among the least litigious of all nations, with Canada ranking first among the number of such cases having been decided. An example of an important decision coming from the Canadian judicial system is the following decision from the Environmental Review Tribunal, whose primary role it is to adjudicate applications and appeals under various environmental and planning statutes in the province of Ontario. This case involved a complaint about a proposed wind facility based on the potential damage to human health, nonhuman animals, and plants in the region of the wind facility. The Tribunal's decision included the major sections reprinted next. Note that the acronym REA stands for "renewal energy approval," the acronym PECFN stands for Prince Edward County Field Naturalists, *and the term post-turbine witness refers to 11 individuals who resided within 2 kilometers of an operating wind turbine project in Ontario. Ellipses note the omission of text or references.*

The issues are:

1. Whether engaging in the Project in accordance with the REA will cause serious harm to human health.
2. Whether engaging in the Project in accordance with the REA will cause serious and irreversible harm to plant life, animal life or the natural environment.
3. If the answer to either Issue 1 (a) or (b) is "yes", whether the Tribunal should revoke the decision of the Director, by order direct the Director to take some action, or alter the decision of the Director. . . .

Conclusion on Issue 1

The evidence in this proceeding did not establish a causal link between wind turbines and either direct or indirect serious harm to human health at the 550 m set-back distance required under this REA.

The evidence in this hearing did not establish that the Ostrander Point Project operating in accordance with the REA will cause serious harm to human health.

For these reasons the Tribunal finds that the appellant has not established that engaging in the Project in accordance with the REA will cause serious harm to human health, and dismisses APPEC's appeal. . . .

Tribunal findings on Sub-Issue 1 (Causation)

The Tribunal finds that it cannot rely on the testimony of the post-turbine witnesses to make the link between their health complaints and the wind turbines. The reasons for this finding include:

- A finding that wind turbine noise causes harm to human health would be a medical conclusion. The panel has no medical expertise and must therefore rely on experts in the field. . . .

[The Tribunal then reviews the specific reasons for reaching this decision.]

In conclusion, the Tribunal finds that, taking the post-turbine witnesses' testimony and all of the expert evidence and Dr. McMurtry's proposed Case Definition together, APPEC has not established that the alleged health effects are caused either by direct exposure to wind turbine noise, or indirectly through some other mechanism. . . .

The evidence in this proceeding did not establish a causal link between wind turbines and either direct or indirect serious harm to human health at the 550 m set-back distance required under this REA.

The evidence in this hearing did not establish that the Ostrander Point Project operating in accordance with the REA will cause serious harm to human health.

For these reasons the Tribunal finds that the appellant has not established that engaging in the Project in accordance with the REA will cause serious harm to human health, and dismisses APPEC's appeal. . . .

Issue No. 2: Whether Engaging in the Project in Accordance with the REA Will Cause Serious and Irreversible Harm to Plant Life, Animal Life or the Natural Environment.

Sub-issue 1: Animal Life

The Tribunal finds that mortality due to roads, brought by increased vehicle traffic, poachers and predators, directly in the habitat of Blanding's turtle, a species that is globally endangered and threatened in Ontario, is serious and irreversible harm to Blanding's turtle at Ostrander Point Crown Land Block that will not be effectively mitigated by the conditions in the REA.

The Tribunal finds that the appellant has not established that engaging in the Project in accordance with the REA will cause serious and irreversible harm to birds or their habitat.

The Tribunal concludes that PECFN has not established that engaging in the Project in accordance with the REA will cause serious and irreversible harm to bats.

The Tribunal finds that PECFN has not established that engaging in the Project in accordance with the REA will cause serious and irreversible harm to Monarch butterflies.

Sub-issue 2: Plant Life

The Tribunal finds that PECFN has not shown that engaging in the Project in accordance with the REA, (i.e., including the minimum mitigation measures outlined in s. I17 of the REA that must be included in a future ARMP), will cause serious and irreversible harm to alvar plants or the alvar ecosystem at the Ostrander Point Crown Land Block.

Source: *Alliance to Protect Prince Edward County v. Director, Ministry of the Environment.* http://www.ert.gov.on.ca/english/decisions/index.htm. Accessed on June 5, 2014. © Queen's Printer for Ontario, 2013.

Wind Energy: Additional Actions Could Help Ensure Effective Use of Federal Financial Support, GAO Report 13–136 (2013)

At the request of the U.S. Congress, the Government Accountability Office (GAO) prepared a report in 2013 on the funding of wind research and development in the United States by the federal government. The major conclusions reported by GAO in that study were as follows:

What GAO Found

GAO identified 82 federal wind-related initiatives, with a variety of key characteristics, implemented by nine agencies in fiscal year 2011. Five agencies—the Departments of Energy (DOE), the Interior, Agriculture (USDA), Commerce, and the Treasury—collectively implemented 73 of the initiatives. The 82 initiatives incurred about $2.9 billion in wind-related obligations and provided estimated wind-related tax subsidies totaling at least $1.1 billion in fiscal year 2011, although complete data on wind-related tax subsidies were not available. Initiatives supporting deployment of wind facilities, such as those financing their construction or use, constituted the majority of initiatives and accounted for nearly all obligations and estimated tax subsidies related to wind in fiscal year 2011. In particular, a tax expenditure and a grant initiative, both administered by Treasury, accounted for nearly all federal financial support for wind energy.

The 82 wind-related initiatives GAO identified were fragmented across agencies, most had overlapping characteristics, and several that financed deployment of wind facilities provided some duplicative financial support. The 82 initiatives were fragmented because they were implemented across nine agencies, and 68 overlapped with at least one other initiative because of shared characteristics. About half of all initiatives reported formal coordination. Such coordination can, in principle, reduce the risk of unnecessary duplication and improve the effectiveness of federal efforts. However, GAO identified 7 initiatives that have provided duplicative support—financial support from multiple initiatives to the same recipient for deployment of a single project. Specifically, wind project developers have in many cases combined the support of more than 1 Treasury initiative and, in some cases, have received additional support from smaller grant or loan guarantee programs at DOE or USDA. GAO also identified 3 other initiatives that did not fund any wind projects in fiscal year 2011 but that could, based on their eligibility criteria, be combined with 1 or more initiatives to provide duplicative support. Of the 10 initiatives, those at Treasury accounted for over 95 percent of the federal financial support for wind in fiscal year 2011.

Agencies implementing the 10 initiatives allocate support to projects on the basis of the initiatives' goals or eligibility criteria, but the extent to which applicant financial need is considered is unclear. DOE and USDA—which have some discretion over the projects they support through their initiatives—allocate support based on projects' ability to meet initiative goals such as reducing emissions or benefitting rural communities, as well as other criteria. Both agencies also consider applicant need for the support of some initiatives, according to officials. However, GAO found that neither agency documents assessments of applicant need; therefore the extent to which they use such assessments to determine how much support to provide is unclear. Unlike DOE and USDA, Treasury generally supports projects based on the tax code's eligibility criteria and does not have discretion to allocate support to projects based on need. While the support of these initiatives may be necessary in many cases for wind projects to be built, because agencies do not document assessments of need, it is unclear, in some cases, if the entire amount of federal support provided was necessary. Federal support in excess of what is needed to induce projects to be built could instead be used to induce other projects to be built or simply withheld, thereby reducing federal expenditures.

Source: "Wind Energy: Additional Actions Could Help Ensure Effective Use of Federal Financial Support." GAO Report 13-136. http://www.gao.gov/assets/660/652957.pdf. Accessed on April 17, 2014.

Wind Technologies Market Report (2013)

The U.S. Department of Energy annually publishes a report on the status of wind technologies in the United States. The report is a very useful source of information about almost every conceivable aspect of the wind energy industry in this country. A summary of the findings in the 2012 report is as follows (detailed discussions of each finding are not included here but are available in the report itself):

Key findings from this year's Wind Technologies Market Report include:

- Wind power additions hit a new record in 2012, with 13.1 GW of new capacity added in the United States and $25 billion invested.
- Wind power represented the largest source of U.S. electric-generating capacity Additions in 2012.
- The United States narrowly regained the lead in annual wind power capacity additions in 2012 but was well behind the market leaders in wind energy penetration.
  - Texas added more new wind power capacity than any other state, while nine states exceed 12% wind energy penetration.
  - No commercial offshore turbines have been commissioned in the United States, but offshore project and policy developments continued in 2012.
  - Data from interconnection queues demonstrate that an enormous amount of wind power capacity is under consideration but that relative interest in wind may be declining.

- The "Big Three" turbine suppliers captured more than 70% of the U.S. market in 2012, yet diversification continues.
- The manufacturing supply chain responded to a record year in wind power capacity additions, but with substantial growing pains.
- Despite challenges, a growing percentage of the equipment used in U.S. wind power projects has been sourced domestically in recent years.
- Although the average nameplate capacity of installed wind turbines declined slightly, the average hub height and rotor diameter continued to increase.
- The project finance environment held steady in 2012.
- Independent power producers remained the dominant owners of wind projects while utilities took a breather in 2012.
- Long-term contracted sales to utilities remained the most common off-take arrangement and have gained ground since the peak of merchant development in 2008/2009.
- Wind turbine prices remained well below levels seen several years ago.
- Reported installed project costs continued to trend lower in 2012.
- Installed costs differed by project size, turbine size, and region.
- Operations and maintenance cost varied by project age and commercial operations date.
- Trends in sample-wide capacity factors were impacted by curtailment and inter-year wind resource variability.
- Average capacity factors for projects built after 2005 have been stagnant: Turbine design changes boosted capacity factors, while project build-out in lower-quality resource areas pushed the other way.

- Regional variations in capacity factor reflect the strength of the wind resource.
- Wind power purchase agreement prices generally have been falling since 2009 and now rival previous lows set a decade ago (despite the trend toward lower-quality wind resource sites).
- Low wholesale electricity prices continued to challenge the relative economics of wind power.
- Short-Term extension of federal incentives for wind energy has helped restart the domestic market.
- State policies help direct the location and amount of wind power development, but current policies cannot support continued growth at recent levels.
- Solid progress on overcoming transmission barriers continued.
- System operators are implementing methods to accommodate increased penetration of wind energy.

Source: "2012 Wind Technologies Market Report. U.S. Department of Energy. http://energy.gov/sites/prod/files/2013/12/f5/2012_wind_technologies_market_report.pdf. Accessed on April 17, 2014.

Offshore Wind Market and Economic Analysis (2013)

Each year, the U.S. Department of Energy commissions a study of the global status of offshore wind generation. The major findings of the 2013 reported are listed here. Detailed discussions of these points constitute the core of the report and can be found in the source cited at the end of the selection.

Section 1. Global Offshore Wind Development Trends

There are approximately 5.3 gigawatts (GW) of offshore wind installations worldwide.

Since the last edition of this report, several potential U.S. offshore wind projects have achieved notable advancements in their development processes.

On the demonstration-project front, the University of Maine, in partnership with the U.S. Department of Energy (DOE), installed the United States' first offshore wind turbine: a 1/8-scale pilot turbine on a floating foundation.

Offshore wind power prices have generally increased over time. For projects installed in 2012 (for which data was available), the average reported capital cost was $5,384/kW.

The average nameplate capacity of offshore wind turbines installed globally each year has grown from 2.9 MW in 2007 to 4.1 MW in 2012.

Globally, offshore wind projects continue to trend further from shore into increasingly deeper waters; parallel increases in turbine sizes and hub heights are contributing to higher reported capacity factors.

Approaches to drivetrain configurations continue to diversify in an effort to improve reliability and reduce exposure to volatile supplies of the rare earth metals required for direct drive generators.

The general trend toward diversification of substructure types also continued in 2012 and 2013, as the industry seeks to address deeper waters, varying seabed conditions, increasing turbines sizes, and the increased severity of wind and wave loading at offshore wind projects.

Few new developments have occurred with regard to vessels, logistics, and the operations and maintenance (O&M) of offshore wind farms since the previous edition of this report.

Section 2. Analysis of Policy Developments

U.S. offshore wind development faces significant challenges: (1) the cost competitiveness of offshore wind energy; (2) a lack of infrastructure such as offshore transmission and purpose-built ports and vessels; and (3) uncertain and lengthy regulatory processes.

For the U.S. to maximize offshore wind development, the most critical need continues to be stimulation of demand through addressing cost competitiveness.

Increased infrastructure is necessary to allow demand to be filled.

Regulatory policies cover three general categories: (a) policies that define the process of obtaining site leases; (b) policies that define the environmental, permitting processes; and (c) policies that regulate environmental and safety compliance of plants in operation.

Section 3. Economic Impacts

Current employment levels could be between 150 and 590 full-time equivalents (FTEs), and current investment could be between $21 million and $159 million.

Section 4. Developments in Relevant Sectors of the Economy

The development of an offshore wind industry in the U.S. will depend on the evolution of other sectors in the economy.

Factors in the power sector that will have the largest impact include natural gas prices and the change in coal-based generation capacity.

Source: "Offshore Wind Market and Economic Analysis." U.S. Department of Energy. http://www1.eere.energy.gov/wind/pdfs/offshore_wind_market_and_economic_analysis_10_2013.pdf. Accessed on April 18, 2014.

Specific Standards for Wind Facilities, State of Oregon (2014)

States have only recently begun to consider the necessity of establishing rules and regulations for the siting of wind power facilities. One of the first states to adopt legislation of this kind was Oregon, whose current regulations on this issue are summarized here.

Specific Standards for Wind Facilities
[Section] 345-024-0010

Public Health and Safety Standards for Wind Energy Facilities

To issue a site certificate for a proposed wind energy facility, the Council must find that the applicant:

(1) Can design, construct and operate the facility to exclude members of the public from close proximity to the turbine blades and electrical equipment.

(2) Can design, construct and operate the facility to preclude structural failure of the tower or blades that could endanger the public safety and to have adequate safety devices and testing procedures designed to warn of impending failure and to minimize the consequences of such failure.

Stat. Authority: ORS 469.470, ORS 469.501
Stat. Implemented: ORS 469.501
[Section] 345-024-0015

Cumulative Effects Standard for Wind Energy Facilities

To issue a site certificate for a proposed wind energy facility, the Council must find that the applicant can design and construct the facility to reduce cumulative adverse environmental effects in the vicinity by practicable measures including, but not limited to, the following:

(1) Using existing roads to provide access to the facility site, or if new roads are needed, minimizing the amount of land used for new roads and locating them to reduce adverse environmental impacts.

(2) Using underground transmission lines and combining transmission routes.

(3) Connecting the facility to existing substations, or if new substations are needed, minimizing the number of new substations.
(4) Designing the facility to reduce the risk of injury to raptors or other vulnerable wildlife in areas near turbines or electrical equipment.
(5) Designing the components of the facility to minimize adverse visual features.
(6) Using the minimum lighting necessary for safety and security purposes and using techniques to prevent casting glare from the site, except as otherwise required by the Federal Aviation Administration or the Oregon Department of Aviation.

Source: Chapter 345: Oregon Administrative Rules, Division 24. http://www.oregon.gov/energy/Siting/docs/rules/div24.pdf. Accessed on March 8, 2014.

Ostrander Point GP Inc. and Another v. Prince Edward County Field Naturalists and Another, 2014 ONSC #974 (2014)

In July 2013, the Ontario (Canada) Environmental Review Tribunal (ERT) issued a ruling revoking a Renewable Energy Approval (REA) that had been issued to Ostrander Point GP for the construction of nine wind turbines on a site in Prince Edward County. The ruling was based solely on the presumption of harm that would be done by the facility to a community of Blanding's turtle, an endangered species living in the area. The ERT rejected appeals by the Prince Edward County Field Naturalists and the Alliance to Protect Prince Edward County claiming that the wind turbines would also cause harm to plants and animals living in the area and to the health of humans living nearby. The ERT's decision was appealed to the Ontario Superior Court of Justice (SCJ), which issued its ruling on February 20, 2014. The case is of special interest because it reached one of the highest judicial levels of any wind energy case anywhere in the world to that date. The SCJ agreed with the ERT's ruling that the wind farm would cause no harm to plants or animals or to human health but overruled its decision on the Blanding's turtle, effectively allowing the Ostrander Point project to proceed as planned. Asterisks note the omission of technical citations.

Conclusion on appeal #1

[91] I have concluded that the Tribunal committed the following errors of law in arriving at its conclusion:

(i) the Tribunal failed to separately identify and explain its reasons for concluding that, if serious harm would result from the Project, that serious harm was irreversible;

(ii) the Tribunal concluded that serious and irreversible harm would be occasioned to Blanding's turtle without any evidence as to the population size affected;

(iii) the Tribunal concluded that serious and irreversible harm would be occasioned to Blanding's turtle arising from road mortality without any evidence as to the current level of vehicular traffic on the Project site or any evidences as to the degree of increase in vehicular traffic arising from the Project;

(iv) the Tribunal failed to give sufficient weight to the existence of the ESA [Endangered Species Act] permit, the conditions attached to that permit, the obligation of the MNR [Ministry of Natural Resources] to monitor and enforce the permit and the fact that the Renewable Energy Approval expressly required Ostrander to comply with the ESA permit;

(v) the Tribunal failed to give a proper opportunity to the parties to address the issue of the appropriate remedy and thereby violated the principles of natural justice and procedural fairness;

(vi) the Tribunal erred in finding that it was not in a position to alter the decision of the Director, or to substitute its opinion for that of the Director.

[92] The Tribunal's decision on serious and irreversible harm was unreasonable for the reasons that I have given. The Tribunal's decision on the appropriate remedy was reached following on a breach of the rules of natural justice by failing to accord procedural fairness to the parties. Consequently, neither of the decisions of the Tribunal can stand. The appeal is therefore allowed and the decision of the Tribunal is set aside. . . .

[98] Moving then to the submissions of PECFN [Prince Edward County Field Naturalists] on the issues of birds and alvar, I should first note that there is agreement between the parties that the standard of review applicable to the Tribunal's decision on this issue is reasonableness.

(i) Birds

[99] The Tribunal concluded that the evidence placed before it did not establish that the Project would cause serious harm to migratory bird habitat. The Tribunal found that the Project "might" cause such harm but that the evidence failed to establish that the Project "will" cause such harm.

[100] In reaching that conclusion, the Tribunal said that the expert witnesses for both sides substantially agreed that a wide variety of species, and large numbers of individual birds, "are found at, or pass through, the Prince Edward County south shore peninsula". The Tribunal found that birds heavily use Ostrander Point and that some species are migratory while others breed in the area. The Tribunal noted that Environment Canada had described the site as "one of the best areas for birds" that it had seen in southern Ontario.

[101] At the same time, the Tribunal noted the mitigation measures for the Project relating to birds including blade feathering and the possible shut down of individual turbines. It also pointed to a radar early detection system that would alert the operators of the turbines to any influx of large groups of birds and to the Project's 200 metre set-back from Lake Ontario. In reaching its conclusion, the Tribunal, correctly in my view, found that the mitigation measures for birds in the Renewable Energy Approval was part of its statutory mandate to consider the operation of the Project in accordance with the Renewable Energy Approval ***.

[102] PECFN asserts that the Tribunal failed to provide an interpretation of the scale that it used to assess whether serious and irreversible harm would be caused to birds by the Project. In failing to do so, PECFN submits that the Tribunal failed to give adequate reasons for its decision and therefore committed an error of law.

[103] I do not agree with PECFN's core submission. While I accept that the reasons of the Tribunal do not have a separate and distinct portion that address the scale issue directly, unlike the situation on the first appeal, the Tribunal's consideration and analysis of this issue is readily apparent from a review of the reasons as a whole. The Tribunal noted that the two experts called by Ostrander, namely Dr. Strickland and Dr. Kerlinger, had expertise in the specific area of wind turbine impacts on bird populations and mortality. The Tribunal preferred the evidence of these two experts over those called by PECFN, as the Tribunal was entitled to do. . . .

[105] It is, in my view, clear from the Tribunal's reasons that, based on the evidence of these two experts, the Tribunal determined that a regional population for birds was the appropriate scale to be used in evaluating whether the Project would cause serious and irreversible harm. The Tribunal implicitly rejected the position of PECFN that the relevant

scale was that of the Project site alone. The Tribunal noted the expert evidence that bird populations are fluid and may extend out for "50 or 100 miles in any direction". The Tribunal also had the evidence of Dr. Kerlinger and Dr. Strickland who made the relative comparison between the small size of the Project (approximately 0.06 km2) and the size of the Important Bird Area of the Prince Edward County South Shore (approximately 279 km2) in terms of evaluating the likely impact of the Project on the relevant bird populations.

[106] It is not therefore a fair criticism of the Tribunal's decision to assert that it failed to identify the scale that it used for determining serious and irreversible harm to birds. What PECFN's complaint boils down to is that the Tribunal did not adopt the scale that PECFN contended that it should. That, however, was a matter for the Tribunal to decide and its conclusion on that point, given the expert evidence, cannot be shown to be unreasonable.

[107] PECFN also asserts that the Tribunal improperly considered the mitigation measures that Ostrander was obliged to put into place to address the possibility of increased bird mortality, as I have mentioned above. PECFN contends that there was a total lack of evidence to support the effectiveness of those mitigation measures. That contention is not borne out by the evidence. There was evidence of the effectiveness of the mitigation measures. The Tribunal reviewed that evidence and it expressly noted that some of those measures, notably the radar system, were controversial and, in the specific case of the radar system, unproven. However, as the Tribunal correctly noted, the statutory onus rests on the person appealing the Renewable Energy Approval. The Tribunal found that PECFN had not met its onus on this issue. Specifically, the Tribunal said:

The Tribunal finds that the PECFN has not proven that the mitigation measures incorporated into the REA regarding birds are so deficient that the Project will cause "serious and irreversible harm".

[108] Given the evidence that was before the Tribunal, its conclusion on this point was, once again, a reasonable one.

(ii) Alvar

[109] The other issue raised was the potential impact of the Project on plant life, namely, alvar. Alvars are described in the Tribunal's reasons, quoting from the Federation of Ontario Naturalists' publication, "The Alvars of Ontario", as "naturally open areas of thin soil over flat limestone or marble rock with trees absent or at least not forming a continuous canopy". The Tribunal noted that alvars are globally imperilled.

[110] The Tribunal had expert evidence that the Ostrander Point area is an alvar or alvar landscape. The Tribunal also noted that the Ostrander Point Crown Land Block within the Ostrander Point area is considered a "Legacy site" by the Department of National Defence. The area had been used by the DND as a bombing, gunnery and rocket range in 1952. It had also apparently been used for tank manoeuvres. The Tribunal found that there had been significant disturbance to the Ostrander Point site in the past, arising from these and other activities, although the precise degree of disturbance was not clear.

[111] On this issue, PECFN submits that the Tribunal's finding, that the Project would not cause irreversible harm to the alvar, was not supported by the evidence. It contends that there was no evidence regarding the extent of the disturbance caused by military activities and therefore the Tribunal improperly concluded that, since the alvar regenerated after that activity, it would likely regenerate after the Project's activity. Indeed, in its oral submissions, PECFN went so far as to suggest that the military activity had not occurred at the Project site but in an area adjacent

to it. PECFN apparently based this latter suggestion on the opinion of its expert, Dr. Catling, who had visited the site and had not observed any disturbance on it consistent with military use.

[112] I do not accept either of PECFN's submissions on this point. . . . The critical issue was that the site had suffered significant disturbance that would have impacted on the alvar but that the alvar had recovered from that disturbance. That reality provided a sufficient foundation for the Tribunal's conclusion that the alvar would likely recover from any disturbance associated with the Project and therefore there was no basis upon which it could conclude that any harm to the alvar was irreversible. More specifically, the Tribunal said:

> The Tribunal finds that the Ostrander Point Crown Land Block has recovered to the status of an important, diverse, self-sustaining alvar, following severe disturbance in the past. This past recovery mitigates against a finding that the harm to plant life in this case will be irreversible.

In reaching its conclusion on this issue, the Tribunal also correctly considered the mitigation requirements contained in the Renewable Energy Approval.

(iii) Conclusion

[115] On appeal #2, I find that the conclusions of the Tribunal on the issue of harm to the birds and harm to the alvar are reasonable ones. They are entirely within the scope of the Tribunal's statutory mandate and they are borne out by the evidence heard and evaluated by the Tribunal. Certainly, none of the issues raised regarding the conclusions reached by the Tribunal rise to the level of an error of law that would give this court jurisdiction to intervene.

[116] The appeal by PECFN regarding the Tribunal's decision on birds and alvar is dismissed.

Appeal #3—serious harm to human health

[117] This appeal can be dealt with more briefly. I will, at the outset, point out that the test under the EPA is different for this issue than it is for the issue regarding harm to plant or animal life. For the purposes of harm to human health, the test only requires a finding of serious harm and not the additional requirement of irreversible harm. I will note that there is again agreement between the parties that the standard of review applicable to the Tribunal's decision on this issue is reasonableness.

[118] The Tribunal made two findings that are central to its conclusion that serious harm to human health had not been demonstrated by APPEC to the level required under s. 145.2.1(2)(b). One of those findings was that the fact witnesses, who gave evidence regarding their perceived effects from exposure to wind turbines, were inherently unreliable. That conclusion drew from the fact that the expert witnesses gave evidence that the subjective recall of individuals regarding health effects has been shown, through scientific studies, generally to be unreliable. The Tribunal went on to give examples of how that general lack of reliability was revealed in this case with respect to the fact witnesses from whom the Tribunal had heard. In at least four instances, fact witnesses had reported health effects or changes that were clearly demonstrated not to be related to wind turbines despite the witnesses' fervent belief that they were.

[119] The other finding of the Tribunal was with respect to APPEC's expert witness, Dr. McMurtry. The Tribunal rejected Dr. McMurtry's evidence largely because it was based on a theory that had yet to be proven. The Tribunal also found Dr. McMurtry's theory was vague with respect to the radius around the wind turbines within which effects will be experienced and that there was no indication of the prevalence of symptoms within exposed persons. The Tribunal ultimately concluded that neither the expert evidence called by APPEC, nor the factual evidence, established, on a

balance of probabilities, that any of the alleged health ill-effects were caused by wind turbines.

[120] APPEC says that the Tribunal erred in this conclusion because it subjected their evidence to a standard of scientific certainty rather than deciding it on a balance of probabilities. I do not agree. In my view, the core problem with APPEC's submission is that it confuses the standard for admissible expert evidence with the standard to be applied in deciding the ultimate issue, that is, whether the test under s. 145.2.1(2) has been met.

[121] All parties agree that the test under s. 145.2.1(2) is to be decided on a balance of probabilities. That is the standard that the Tribunal applied in reaching its decision. However, there is a higher standard that is required before a fact finder can admit and rely on expert evidence, especially when that expert evidence promotes a novel scientific theory. The balance of probabilities onus of proof does not apply to the reliability of a scientific theory. For a court to conclude that a novel scientific theory is reliable, there must be more than a finding that the theory is more probable or more likely than not. Rather, it requires the fact finder to be satisfied that the theory is, in fact, a reliable one. . . .

[122] It is not sufficient for the purposes of relying on a novel scientific theory to simply conclude that the theory may be correct. In that situation, the theory will not have crossed the threshold of reliability for the purpose of establishing the necessary causal link between the activity in issue and the consequences said to arise from that activity. Rather, the party attempting to rely on a *novel* scientific theory must first establish threshold reliability before the fact finder may consider it. . . .

Summary

[129] The appeal by Ostrander and the Director from the decision of the Tribunal is allowed and the Tribunal's decision to

revoke the Director's decision to grant the Renewable Energy Approval is set aside. The appeal by PECFN and the appeal by APPEC are both dismissed.

Source: Ostrander Point GP Inc. and another v. Prince Edward County Field Naturalists and another. 2014 ONSC #974. http://canlii.ca/t/g360c. Accessed on April 20, 2014.

6 Resources for Further Research

Introduction

The history and technology of wind power, as well as the social, political, economic, and other issues associated with that technology, have been the subject of an extensive array of books, articles, reports, and Web sites. This chapter presents an annotated bibliography of a selection of these resources.

Books

Ahrens, Uwe, Moritz Diehl, and Roland Schmehl. 2014. *Airborne Wind Energy.* Berlin; Heidelberg: Springer.

This book is divided into five sections, dealing with fundamentals; system modeling, optimization, and control; analysis of flexible kite dynamics; implemented concerns; and component design. The early parts of the book are accessible to the general reader, whereas the latter parts are more appropriate for specialists in the field.

Anaya-Lara, Olimpo, David Campos-Gaona, Edgar Moreno-Goytia, and Grain Adam. 2014. *Offshore Wind Energy Generation: Control, Protection, and Integration to Electrical Systems.* Hoboken, NJ: Wiley.

This important reference provides a complete and technical discussion of the major issues surrounding the design,

---

The average American home would require a wind turbine with a 5-KW generating capacity in order to meet all of its electrical needs. (National Renewable Energy Laboratory)

development, and operation of offshore wind facilities, with special attention to the technical issues involved in the operation of such systems.

Bennett, Martha C. 2013. *Wind Power Trends and Offshore Wind Projects in the United States.* New York: Nova Publishers.

This book reviews the development of wind power in the United States, especially since 2010, with an analysis of the governmental actions that have affected its growth. The book then explores the potential future of wind power, with possible governmental actions that may influence future development.

Blackwell, Ben. 2014. *Wind Power: The Struggle for Control of a New Global Industry.* London: Routledge.

This book is based on the fact that wind power has grown from a niche technology to one of the fastest-growing and potentially most important energy sources in the twenty-first century. The author examines the way in which nations, corporations, and individuals are involved in the changing character of wind technology.

Bostan, Ion, et al. 2013. *Resilient Energy Systems: Renewables: Wind, Solar, Hydro.* Dordrecht; New York: Springer.

This book provides an excellent general introduction to a variety of renewable energy sources, including wind. The wind section provides a solid, basic introduction to the topic, followed by a somewhat more detailed and advanced analysis of some of the primary issues involved in the development of wind energy facilities.

Busby, Rebecca L. 2012. *Wind Power: The Industry Grows Up.* Tulsa, OK: PennWell Corp.

The author provides a well-written, general introduction to the history and current status of the wind energy industry, suitable for the general reader.

Chauhan, Amit. 2014. *Vertical Axis Wind Turbine Analysis of Airfoils, Present Wind Energy Scenario.* Saarbrücken, Germany: LAP LAMBERT Academic Publishing.

Somewhat less widely discussed and used on a commercial scale, vertical-axis wind turbines are the subject of this book, which provides a general introduction to the topic and a review of the technology of this type of wind device.

Chiras, Daniel D., Mick Sagrillo, and Ian Woofenden. 2009. *Wind Power Basics.* Gabriola Island, BC, Canada: New Society Publishers.

This book is directed at individuals and small businesses who are increasingly concerned about the rising costs of energy consumption and are looking for alternatives to traditional utility-provided energy. The book describes the nature, construction, and operation of small-turbine systems.

Clark, R. Nolan. 2014. *Small Wind: Planning and Building Successful Installations.* Waltham, MA: Academic Press/Elsevier.

This book is designed for individuals who are interested in building small (less than 100 kW) wind systems for personal or small business use. Some of the chapters in the book deal with topics such as various aspects of site evaluation and need evaluation, wind turbine components, towers and foundations, permitting, electrical interconnections, economic considerations, distributed systems, and the future of small wind.

Emeis, Stefan. 2013. *Wind Energy Meteorology: Atmospheric Physics for Wind Power Generation.* Berlin; New York: Springer.

The purpose of this book is to provide readers with the background needed to investigate meteorological factors related to the operation of wind turbines so that those factors can be taken into account during the design and operations of wind facilities.

Energy: Wind. 2010. Alexandria, VA: TheCapitol.Net.

This publication is written by an organization previously known as Congressional Quarterly Executive Conferences to provide exhaustive information on topics of current political and social interest. The chapters in the book provide excellent and thorough reviews of important aspects of wind energy development and use, including history of wind energy, how wind turbines work, types of wind turbines, wind power generation in the United States, distributed wind energy issues, key data and statistics about wind energy, and the Wind and Water Power Program.

Etzler, J.A. 1836. *The Paradise within the Reach of All Men, without Labour, by Powers of Nature and Machinery: An Address to All Intelligent Men.* London: J. Brooks. Available online at http://books.google.com/books?id=FHUSAAAAIAAJ&printsec=frontcover&dq=john+etzler+paradise&ei=8pc2Soe7OKrYygTK3aBN#v=onepage&q=john%20etzler%20paradise&f=false. Accessed on April 18, 2014.

This book is a fascinating bit of history in which the author describes the many benefits that can accrue to humans by the development of wind power (along with other unconventional sources of energy).

Galbraith, Kate, and Asher Price. 2013. *The Great Texas Wind Rush: How George Bush, Ann Richards, and a Bunch of Tinkerers Helped the Oil and Gas State Win the Race to Wind Power.* Austin: University of Texas Press.

As the title says, the authors explain how an unlikely consortium of political leaders and technological experts brought wind power to the state of Texas, making it today the nation's largest producer of wind-powered electricity.

Gillis, Christopher. 2008. *Wind Power.* Atglen, PA: Schiffer Publishing, Ltd.

This book provides a general introduction to the history of wind power and its growing use as a major source of electrical energy in the late twentieth century, with a consideration of its potential value as an alternative source of energy in the future.

Hau, Erich. 2013. *Wind Turbines: Fundamentals, Technologies, Application, Economics,* 3rd ed. Berlin; New York: Springer, 2013.

This book is designed for specialists in the field of wind turbine design and operation, but it contains material that is accessible to, and of interest for, individuals without specialized background, including historical chapters on the development of wind machines and wind turbines.

Johnson, Rody. 2014. *Chasing the Wind: Inside the Alternative Energy Battle.* Knoxville: The University of Tennessee Press.

The author provides an excellent overview of the development of wind power as a possible factor in the world's efforts to solve future energy shortages. He then places that issue within the context of an ongoing battle over the installation of a wind farm in Greenbrier County, West Virginia, showing how many of the elements involved in the development of wind power actually play out within an existing social and political structure in a rural American setting.

Kishore, Ravi, and Shashank Priya. 2014. *Wind Energy Harvesting: Micro-to-Small Scale Turbines.* Boston: De Gruyter.

The authors provide a general introduction to the topic of small wind systems with a discussion of the technical issues and problems involved with the design, construction, and operation of such systems.

Kothari, D.P., Nagpur S. Umashankar, and Shikha Singh. 2014. *Wind Energy Systems and Applications.* Oxford, UK: Alpha Science International.

This book is intended for specialists in the field and provides detailed analysis of the technology of wind turbine systems and their uses in a variety of applications.

Letcher, Trevor. 2014. *Future Energy: Improved, Sustainable and Clean Options for Our Planet,* 2nd ed. London: Elsevier.

This book provides an excellent introduction for the general public to the topic of all forms of energy in the world's future energy equation. Chapter 15 focuses on wind power and deals with topics such as wind resources, wind turbine technology, environmental impacts of wind power, and potential future developments.

Lewis, Joanna I. 2013. *China's Wind Power Industry and the Global Transition to a Low-Carbon Economy.* New York: Columbia University Press.

China is now the world's largest producer of wind powered energy, so an understanding of that country's industry is essential to an understanding of the future of wind power worldwide. This excellent book reviews all aspects of the Chinese wind power industry, from its origin and growth to its present status to the role played by the interactions between Chinese and non-Chinese governments and industrial organizations around the world.

Maegaard, Preben, Anna Krenz, and Wolfgang Palz, eds. 2014. *Wind Power for the World: The Rise of Modern Wind Energy,* vol. 1. Singapore: Pan Stanford Publishing.

This book is one in a series of five volumes on alternative energy. It presents a detailed and exhaustive review of the history of early modern wind power production. It focuses primarily on the Danish experience but also reviews and discusses developments in Germany, California, and China.

Maegaard, Preben, Anna Krenz, and Wolfgang Palz, eds. 2014. *Wind Power for the World:* International Reviews and Developments, vol. 2. Singapore: Pan Stanford Publishing.

This book is one in a series of five volumes on alternative energy. Following an introductory section on the history and development of wind power worldwide, the book's remaining chapters present detailed descriptions of wind power programs in a host of countries around the world, from China, India, and South Korea, to Brazil, Chile, and Cuba, to Poland, Russia, and Ukraine. An excellent overview of the status of wind power worldwide as of the book's publication date.

Musgrove, Peter. 2010. *Wind Power.* Cambridge, UK: Cambridge University Press.

The author presents a detailed general overview of the history and development of wind power in human society, with special attention to its potential as an alternative to fossil fuels in the planet's future energy equation. A good general introduction to the topic.

Nakaya, Andrea C. 2013. *What Is the Future of Wind Power?* San Diego, CA: ReferencePoint Press.

This book is one in the company's series on future of renewable energy. As with other books in the series, chapters are devoted to a variety of aspects about the topic, such as the pros and cons of wind energy, environmental effects, the ability of the technology to replace fossil fuel supplies, and the nature of government support for wind energy.

Nelson, Vaughn. 2014. *Wind Energy: Renewable Energy and the Environment,* 2nd ed. Boca Raton, FL: CRC Press.

This book offers an excellent general introduction to the subject of wind energy, including chapters on the history of wind power and the fundamentals of energy, followed by detailed discussions of the basic features of a wind power system, such as wind characteristics, wind resources, wind turbines, economic issues, and technical issues in the design and use of wind devices.

Paradalos, Panos, et al., eds. 2013. *Handbook of Wind Power Systems.* Berlin; Heidelberg: Springer-Verlag.

This book includes chapters on a large number of technical issues involved in the planning, construction, and operation of wind power systems, such as optimization of wind power generation, grid integration, modeling and control of wind power generation systems, and novel wind power systems.

Pasqualetti, Martin J., Robert Richter, and Paul Gipe. 2004. "History of Wind Energy." In Cutler J. Cleveland, ed., *Encyclopedia of Energy,* vol. 6. San Diego: Academic Press, 419–433.

The authors provide a general history of wind energy.

Pierpont, Nina. 2009. *Wind Turbine Syndrome: A Report on a Natural Experiment.* Santa Fe, NM: K-Selected Books.

This book is one of the most controversial pieces of writing on wind turbine effects of anything currently available. The author argues that wind turbines are responsible for a wide variety of physical and health effects that are largely being ignored by wind power supporters and the medical community.

Pinson, Pierre, and Henrik Madsen. 2013. *Forecasting Wind Power Generation: From Statistical Framework to Practical Aspects.* London: Routledge.

Wind energy is a unique type of energy resource in that it depends very heavily on a single natural event: the wind. Successful design, construction, and operation of a wind facility, therefore, depend on knowing to the extent possible what weather conditions are likely to be at a given location at various times of the year. This book provides a theoretical and practical introduction to the solution of this problem.

Rekioua, Ziani Djamila. 2014. *Wind Power Electric Systems: Modeling, Simulation and Control.* Berlin; New York: Springer.

This text is designed for students in the field of wind technology, providing the technical background for dealing with fundamental problems in the field.

Righter, Robert W. 1996. *Wind Energy in America: A History.* Norman: University of Oklahoma Press.

The author provides a superb history of wind power, beginning in ancient times but focusing especially on its development, production, and use in the United States. He also reviews some of the issues involved in support of, and opposition to, the expanded use of wind power in modern society.

Rule, Troy A. 2014. *Solar, Wind and Land: Conflicts in Renewable Energy Development.* London: Routledge.

The author of this work analyzes and discusses issues of land use associated with the siting and operation of solar and wind facilities, primarily in the United States.

Schaffarczyk, Alois, ed. 2014. *Understanding Wind Power Technology: Theory, Deployment and Optimisation.* New York: Wiley.

This textbook is designed for students majoring in the field of wind engineering, providing a broad background in the main principles of that field.

Shogren, Jason F., 2013. *Encyclopedia of Energy, Natural Resource, and Environmental Economics.* Amsterdam; San Diego, CA: Elsevier Academic Press.

This superb reference work has a large number of articles dealing with wind power in general and in many specific aspects of that topic, such as wind power technology, the economics of wind power, the environmental impacts of wind power, and the characteristics and effects of onshore and offshore wind farms.

Signorelli, Vincent, ed. 2014. *Impact of Wind Energy Facilities on Residential Property Values.* New York: Nova Publishers.

One of the most common complaints about wind farms is the deleterious effects they have on property values of adjacent lands. Although a number of small studies have been done on this topic, this book presents what is probably the most complete analysis of the relationship of wind farms and property values currently available.

Swift, Andrew. 2014. *Wind Energy Impacts A Comparison of Various Sources of Electricity.* New York: John Wiley & Sons.

The first few chapters of this book deal with the history of wind power. The core of the book then focuses on comparisons between wind power and other energy sources for the generation of electricity.

Szarka, Joseph, et al. 2012. *Learning from Wind Power: Governance, Societal and Policy Perspectives on Sustainable Energy.* Houndmills, Hampshire, UK; New York: Palgrave Macmillan.

Developing wind power systems involves a great deal more than technical knowledge and skills. Governmental policies, economic conditions, social settings, and a host of other factors are involved in the successful siting and operation of a wind facility. This book takes the position that the process through which wind energy development has gone in the last half century provides useful lessons for the development of other forms of alternative and renewable energy, the topic of the essays that make up the book.

Urley, Martin W. 2013. *Wind Power Market and Economic Trends.* New York: Nova Science Publishers.

This book focuses on the economic potential for distributed farm systems both in the United States and worldwide.

Warburg, Philip. 2014. *Harvest the Wind: America's Journey to Jobs, Energy Independence, and Climate Stability.* Boston: Beacon Press.

The author provides an excellent, comprehensive, and easy-to-read review of the history of wind energy and its growing importance in the future energy equation of the United States and the rest of the world. Chapters deal with specific topics such as the growth of wind power in China, environmental problems posed by wind turbines and wind farms, and community resistance to the development of wind facilities.

Woolfenden, Ian. 2009. *Wind Power for Dummies.* Hoboken, NJ: John Wiley and Sons.

This book provides a brief general overview of wind power but focuses primarily on instructions for the construction of a wind power system at one's own home or business. As with other books in this series, the presentation is at the simplest possible level.

Articles

A number of peer-reviewed journals deal with a variety of wind-related issues, most on a relatively technical level. These journals include *Journal of Wind Energy* (ISSN: 2356-7732, 2314-6249); *Journal of Wind Engineering and Industrial Aerodynamics* (ISSN: 01676105); *Wind and Structures, An International* Journal (ISSN: 12266116); and *Wind Energy* (ISSN: 10954244, 10991824).

Adams, Amanda S., and David W Keith. 2013. "Are Global Wind Power Resource Estimates Overstated?" *Environmental Research Letters.* 8(1): 1–9. Also available online at http://iopscience.iop.org/1748-9326/8/1/015021/pdf/1748-9326_8_1_015021.pdf. Accessed on April 21, 2014.

The authors say the answer to this question is "yes" because researchers tend to ignore the "shadow effect" produced when one turbine interferes with the air stream available to other turbines in its area.

Adaramolaa, Muyiwa S., Martin Agelin-Chaabb, and Samuel S. Paulc. 2014. "Assessment of Wind Power Generation along the Coast of Ghana." *Energy Conversion and Management*. 77: 61–69.

This paper reflects a line of research increasingly common within the academic world, in which scientists attempt to determine the economic viability of establishing wind power facilities at some specific geographical location, in this case along the coast of the African nation of Ghana, with estimates as to the actual costs of such a facility.

Allen, Patricia, and Tim Chatterton. 2013. "Carbon Reduction Scenarios for 2050: An Explorative Analysis of Public Preferences." *Energy Policy*. 63: 796–808.

This study is based on the realization that nations are likely to be adopting methods for reducing the amount of carbon dioxide produced during energy use over the next three decades. The question that might be asked, then, is what options the general public prefers in making this conversion to lower-carbon-emission technologies. This study reports on the answer to that question given by 10,983 respondents in the United Kingdom.

Archera, Cristina L., Luca Delle Monacheb, and Daran L. Rifec. 2014. "Airborne Wind Energy: Optimal Locations and Variability." *Renewable Energy*. 64: 180–186.

The authors review data collected between 1985 and 2005 to determine global locations at which above-ground wind characteristics could be regarded as favorable for the development of high-altitude wind turbine systems. They find that favorable locations are "more ubiquitous than previously thought and can have extraordinarily high wind power densities."

Biello, David. 2012. "Kinetic Kite: An Airborne Wind Turbine Turns Sea Breezes into Electricity." *Scientific American.* 307(4): 66–67.

The author briefly describes the history of airborne wind turbines and discusses one of the current experimental models of the system.

Brown, Jason P., et al. 2012. "Ex Post Analysis of Economic Impacts from Wind Power Development in U.S. Counties." *Energy Economics.* 34(6): 1743–1754.

The authors investigate the question as to the economic impacts on counties where wind power facilities have been installed. They conclude that such counties experience an increase in average personal income of $11,000 per year and an average increase of 0.5 jobs as a result of wind operations.

Delucchi, Mark A., and Mark Z. Jacobson. 2011. "Providing All Global Energy with Wind, Water, and Solar Power, Part II: Reliability, System and Transmission Costs, and Policies." *Energy Policy.* 39(3): 1170–1190.

This paper is the second of two articles by Jacobson and Delucchi explaining how humans can convert from fossil fuels to renewable energy sources in such a way as to be able to meet all energy demands by the year 2030. For the first paper in the series, see Jacobson and Delucchi (2011).

Dvorak, Michael J., et al. 2013. "US East Coast Offshore Wind Energy Resources and Their Relationship to Peak-Time Electricity Demand." *Wind Energy.* 16: 977–997.

The authors attempt to identify the most favorable locations along the East Coast of the United States for the installation of wind turbines. They suggest that the proper placement of turbines in these locations has the theoretical potential to supply about one-third of the total electrical demand in coastal states from Florida to Maine, with the exception of special periods of peak demand.

Eichorn, Marcus, et al. 2012. "Model-Based Estimation of Collision Risks of Predatory Birds with Wind Turbines." *Ecology and Society*. 17(2): 1. Available online at http://www.ecologyandsociety.org/vol17/iss2/art1/. Accessed on April 22, 2014.

The authors of this paper attempt to develop a model for the threat posed by wind systems to various raptor species, with the objective of offering guidance for the design and siting of such systems that will reduce the mortality of raptors.

Fast, Stewart. 2013. "Social Acceptance of Renewable Energy: Trends, Concepts, and Geographies." *Geography Compass*. 7(12): 853–866.

The author reviews a number of public opinion surveys about attitudes toward various forms of renewable energy and finds a changing focus from general acceptance or rejection to more subtle views on the possible negative effects wind farms may have on individuals and communities.

Hammara, Linus, Andreas Wikströmb, and Sverker Molandera. 2014. "Assessing Ecological Risks of Offshore Wind Power on Kattegat Cod." *Renewable Energy*. 66: 414–424.

One of the crucial ongoing questions in the development of wind energy is whether and, if so, how wind energy projects may affect organisms in the natural environment, especially those that may be endangered or threatened. This study is one of the many examples of studies that have been conducted to answer that question. The study shows that that "answer" is often more complicated than a simple "yes" or "no" with regard to harm.

Henriksen, Lars. 2014. "Wind Energy Literature Survey No. 32." *Wind Energy.* 17(8): 1297–1300.

Wind Energy periodically provides a survey of important articles about wind energy from its own issues and from a number of other wind publications. These surveys go back many years and are a very useful bibliographic resource for interested readers. The specific citation given here is for the most recent survey available as of late 2014.

Hernández-Escobedo, Q., et al. 2014. "Wind Energy Resource in Northern Mexico." *Renewable and Sustainable Energy Reviews.* 32: 890–914.

A type of research paper fairly commonly found in wind energy journals today is one that deals with the needs for, and potential of, wind farm facilities in various nations or regions. This paper is cited here is an example of that type. It is possible to find similar articles for a number of regions and countries, many in this journal.

Hitaj, Claudia. 2013. "Wind Power Development in the United States." *Journal of Environmental Economics and Management.* 65(3): 394–410.

The author uses sophisticated tools of analysis to analyze the factors involved in the development of wind power systems in the United States, focusing in particular on the effects of tax credits and other governmental financial incentives.

Jacobson, Mark Z., and Mark A. Delucchi. 2009. "A Path to Sustainable Energy by 2030." *Scientific American.* 301(5): 58–65.

The authors describe a plant by which the world can produce all the energy it needs from renewable resources by the year 2030. The plan includes the construction of 3.8 million wind turbines and 90,000 solar plants. The article suggests that nontechnical issues, such as political factors and the will of the population, are likely to be greater hindrances to the achievement of this goal than are technological challenges. The journal articles on which this article is based can be found at Jacobson and Delucchi (2011) and Delucchi and Jacobson (2011).

Jacobson, Mark Z., and Mark A. Delucchi. 2011. "Providing All Global Energy with Wind, Water, and Solar Power, Part I: Technologies, Energy Resources, Quantities and Areas of Infrastructure, and Materials." *Energy Policy.* 39(3): 1154–1169.

This paper is the first of two articles by Jacobson and Delucchi explaining how humans can convert from fossil fuels to renewable energy sources in such a way as to be able to meet all energy demands by the year 2030. See Delucchi and Jacobson (2011) for the second paper in this series.

Jacobsson, Staffan, and Kersti Karltorp. 2013. "Mechanisms Blocking the Dynamics of the European Offshore Wind Energy Innovation System—Challenges for Policy Intervention? *Energy Policy.* 63(12): 1182–1195.

The authors suggest that greater reliance on alternative forms of energy, such as wind energy, is inevitable in Europe in coming decades. They explore some of the policy issues that must be resolved in dealing with this expanded role of wind energy in national energy equations.

Knopper, Loren D., and Christopher A. Ollson. 2011. "Health Effects and Wind Turbines: A Review of the Literature." *Environmental Health.* 10: 78. Available online at http://www.ehjournal.net/content/10/1/78. Accessed on April 22, 2014.

The authors conclude that published research falls into two general categories: that which involves actual physical effects resulting from wind turbine noise and that which includes emotional or psychological responses, such as annoyance. They conclude that support for the first type of effect is largely absent, whereas the second effect seems relatively well confirmed. They conclude that "further research into the effects of wind turbines (and environmental changes) on human health, emotional and physical, is warranted."

Loss, Scott R., Tom Will, and Peter P. Marra. 2013. "Estimates of Bird Collision Mortality at Wind Facilities in the Contiguous United States." *Biological Conservation.* 168: 201–209.

How many birds are killed in collisions with wind turbines? A number of researchers have designed studies to answer that question, with a wide range of estimates. This study is one of the most recent of those studies. It includes a summary of some earlier studies of the same type, with a discussion of reasons for the range of estimates provided by these studies.

Loyd, M. L. 1980. "Crosswind Kite Power (For Large-Scale Wind Power Production)." *Journal of Energy.* 4(3): 106–111.

This paper is of considerable historical importance because it is generally regarded as the first thoughtful analysis of the technology of high-altitude wind turbines, also known as airborne wind energy, a technology that is just beginning to "take off" in the second decade of the twenty-first century.

Lund, Peter D. 2014. "How Fast Can Businesses in the New Energy Sector Grow? An Analysis of Critical Factors." *Renewable Energy.* 66(1): 33–40.

The author takes note of the fact that demand for and consumption of wind energy has been decreasing in recent years, and he asks which factors are involved in reversing that course and making wind energy more competitive in the overall energy market. He concludes that reducing the cost of wind energy is the primary factor in this equation and suggests ways in which this change can be accomplished in today's economy.

Makridis, Christos. 2013. "Offshore Wind Power Resource Availability and Prospects: A Global Approach." *Environmental Science & Policy.* 33(11): 28–40.

The author reviews previous studies on the probable availability of wind farm siting locations around the world. He then develops a model for making his own assessment of the global locations that are likely to be most favorable for the siting of such facilities, given a variety of social, economic, and technical factors.

McCubbin, Donald, and Benjamin K. Sovacool. 2013. "Quantifying the Health and Environmental Benefits of Wind Power to Natural Gas." *Energy Policy.* 53(4): 429–441.

Many experts on energy issues point to natural gas as the next fossil fuel of significant importance in solving the world's energy problems. The purpose of this article is to attempt to find a precise comparison between benefits and risks to human health and to the natural environment resulting from an increased use of natural gas compared to the use of renewable fuels.

Nordman, Erik E. 2014. "Energy Transitions in Kenya's Tea Sector: A Wind Energy Assessment." *Renewable Energy.* 68: 505–514.

This article is of special interest because it explores the possible application of wind power in an area where the technology is essentially absent at the present time. The author points out that tea farming is an important commercial activity in Kenya, employing about a half million workers, but tea factories are often located at a significant distance from energy sources on which they depend. Wind power may provide a solution to this problem.

Nugent, Daniel, and Benjamin K. Sovacool. 2014. "Assessing the Lifecycle Greenhouse Gas Emissions from Solar PV and Wind Energy: A Critical Meta-survey." *Energy Policy.* 65(5): 229–244.

Renewable energy sources, such as solar and wind power, are often touted because they are "emissions free," posing no threat to the natural environment. They are generally seen as contributing little or nothing to global climate change issues. Nugent and Sovacool examine this issue throughout the life cycle of both solar and wind technologies and find that this argument is not entirely true. They suggest that their analysis can be used to modify the construction of solar and wind projects to reduce the carbon footprint through which they contribute to global climate change.

Perveen, Rehana, Nand Kishor, and Soumya R. Mohanty. 2014. "Off-Shore Wind Farm Development: Present Status and Challenges." *Renewable and Sustainable Energy Reviews.* 29(1): 780–792.

This paper provides a general overview of the current status of offshore wind facilities and some of the economic, technical, political, and other challenges such facilities face.

Rajewski, Daniel A., et al. 2013. "Crop Wind Energy Experiment (CWEX): Observations of Surface-Layer, Boundary Layer, and Mesoscale Interactions with a Wind Farm." *Bulletin of the American Meteorological Society.* 94(5): 655–672.

Researchers studied the way in which wind turbines changed air and heat flow over and through the land on a conventional agricultural farm and found a variety of somewhat unexpected results, some of which led to the conclusion that the presence of a wind farm might actually improve the growing conditions for crops on the land.

Ramos, V., and G. Iglesias. 2014. "Wind Power Viability on a Small Island." *International Journal of Green Energy.* 11(7): 741–760.

The authors explore the possibility of using wind power to supply all the energy needs for a small island in the Atlantic Ocean and find that the installation of six wind turbines would be sufficient to meet all the electrical needs of the island's population for at least twenty-five years into the future.

Salo, Olli, and Sanna Syri. 2014. "What Economic Support Is Needed for Arctic Offshore Wind Power?" *Renewable and Sustainable Energy Reviews.* 31(3): 343–352.

As interest in offshore wind farms grows, new technical, economic, political, social, and other questions arise about the siting and operation of such facilities. This article explores the question as to what those issues might be as plans are being made to install wind farms in very cold environments, such as in offshore waters along the coast of Finland.

Sun, Xiaojing, and Diangui Huanga. 2014. "An Explosive Growth of Wind Power in China." *International Journal of Green Energy.* 11(8): 849–860.

The authors point out that the development of wind power in China has occurred at an explosive rate in the last few decades. They provide a review of the history of wind energy in China, summarize its current status, and discuss the major challenges facing the Chinese wind industry in the future.

Tabassum-Abbasi, M. Premalatha, Tasneem Abbasi, and S.A. Abbasi. 2014. "Wind Energy: Increasing Deployment, Rising Environmental Concerns." *Renewable and Sustainable Energy Reviews.* 31: 270–288.

The authors note that wind power, of all renewable energy sources except perhaps solar, is receiving the greatest attention in recent years as a potential substitute for some of the energy currently being derived from fossil fuels. They also point out that this increased interest in wind power has caused correspondingly more interest in potential environmental problems caused by the technology. This paper provides a review of those environmental concerns.

Tan, Zhongfu, et al. 2014. "Potential and Policy Issues for Sustainable Development of Wind Power in China." *Journal of Modern Power Systems and Clean Energy.* 1(3): 204–215.

The authors of this article examine the current status of wind power programs in China, with consideration of the natural resources available, current economic policies, and legal and regulatory issues. Given this information, they then explore the future potential and possible future directions for the further development of wind power in China.

Timilsina, Govinda R., G. Cornelis van Kooten, and Patrick A Narbel. 2013. "Global Wind Power Development: Economics and Policies." *Energy Policy.* 61(6): 642–652.

The authors explore the factors that have thus far limited wind power facilities from reaching their maximum potential in producing energy for the world's nations. They suggest that the technical issue of intermittency, the unreliability of wind energy, is likely to continue to be its major limiting factor.

Tuohy, Aidan, Ben Kaun, and Robert Entriken. 2013. "Storage and Demand-Side Options for Integrating Wind Power." *Wiley Interdisciplinary Reviews: Energy and Environment.* 3(1): 93–109.

The authors consider the technical and other issues associated with the use of wind power in areas where wind is an unpredictable, variable, and unreliable factor and the ways in which technology and policy can be used to deal with this issue.

Wu, Jie, Zhi-Xin Wang, and Guo-Qiang Wang. 2014. "The Key Technologies and Development of Offshore Wind Farm in China." *Renewable and Sustainable Energy Reviews.* 34(6): 453–462.

The authors explain the reason that wind energy is beginning to play such a critical role in the Chinese energy equation and how existing wind farms have come to fruition. They also discuss some issues involved in the further future expansion of wind energy in China.

Zhang, Guotao, and Xinhua Wan. 2014. "A Wind-Hydrogen Energy Storage System Model for Massive Wind Energy Curtailment." *International Journal of Hydrogen Energy.* 39(3): 1243–1252.

The authors point out that interruption of the flow of energy from wind farms is becoming an increasingly important problem as the number of such farms increases. They describe a system for dealing with this problem that involves the conversion of excess wind energy to hydrogen gas, which can then be stored until it is needed as an energy source.

Zheng, Chong Wei, and Jing Pan. 2014. "Assessment of the Global Ocean Wind Energy Resource." *Renewable and Sustainable Energy Reviews.* 33: 382–391.

The authors use field data collected between 1988 and 2011 to derive an estimate of the energy available from wind on the open seas. They prepare a detailed map of such wind resources on a 0.25 × 0.225-degree grid.

Zhou, Liming, et al. 2012. "Impacts of Wind Farms on Land Surface Temperature." *Nature Climate Change.* 2(7): 539–543.

Researchers study the effects of wind turbine operation on the land on which they are sited and find that the turbines produce significant changes in ground temperature and other physical characteristics. These changes may conceivably have effects on crops grown in the area. For additional papers on this general topic, see the senior author's home page at http://www.atmos.illinois.edu/~sbroy/papers.htm.

Reports

Albrecht, Christina, et al. "The Impact of Offshore Wind Energy on Tourism." http://www.offshore-stiftung.com/60005/Uploaded/Offshore_Stiftung%7C2013_04SBO_SOW_tourism_study_final_web.pdf. Accessed on April 22, 2014.

This report was prepared for a consortium of ten countries bordering on the Baltic Sea interested in expanding wind power presence in the region. The purpose of the report was to determine how the presence of offshore wind farms might be either beneficial or harmful to tourism in the area. It found that some of the problems that had to be overcome included questions about the appropriate use of land and water resources, potential flicker effects, and risks of ship collisions. Benefits provided by the wind farms included fascination with technology, contribution of wind power to protection of the natural environment, and general attractiveness of the region.

Arnett, Edward B., et al. "Impacts of Wind Energy Facilities on Wildlife and Wildlife Habitat." The Wildlife Society. Technical Review 07–2. http://wildlife.org/documents/technical-reviews/docs/Wind07-2.pdf. Accessed on April 22, 2014.

Although somewhat dated (2007), this report provides one of the best general summaries of the risks posed to birds and other wildlife by wind facilities. The report includes chapters on federal and state regulations, mortality rates from collisions, disturbance of habitats by wind turbines, offshore issues, and research needs.

Bolinger, Mark. 2013. "Revisiting the Long-Term Hedge Value of Wind Power in an Era of Low Natural Gas Prices." Lawrence Berkeley National Laboratory. http://emp.lbl.gov/sites/all/files/lbnl-6103e.pdf. Accessed on April 22, 2014.

This report was inspired by the fact that recent developments in the collection of natural gas (by processing such as fracking) have greatly reduced the availability and price of that natural resource. That development, in turn, has made it economically more difficult for wind power to compete in the marketplace with natural gas. Bolinger examines the question of "the degree to which wind power can still serve as a cost-effective hedge against rising natural gas prices, given the significant reduction in gas prices in recent years, coupled with expectations that prices will remain low for many years to come."

Ellison, Laura E. "Bats and Wind Energy—A Literature Synthesis and Annotated Bibliography." Washington, DC: U.S. Department of the Interior; U.S. Geological Survey. http://pubs.usgs.gov/of/2012/1110/OF12-1110.pdf. Accessed on April 22, 2014.

This document is arguably the most complete review of studies on the association between wind power generation and bat mortality available at the time of its publication (2012). In addition to detailed annotations of some of the most important studies, the author provides an extensive additional list of books, articles, theses, and other documents dealing with the topic.

Holburn, Guy, et al. 2013. *Wind Energy in Canada: A Survey of the Policy Environment.* London, ON: Ivey Energy Policy and Management Centre, Western University.

This publication reports on a research study conducted in 2012 to determine the attitudes of private industry to the development of wind energy systems in each of the country's provinces.

Howe, Brian. 2010. Low Frequency Noise and Infrasound Associated with Wind Turbine Generator Systems: A Literature Review." Toronto: Ontario Ministry of the Environment. Available online at https://ia601204.us.archive.org/20/items/stdprod092086.ome/stdprod092086.pdf. Accessed on April 22, 2014.

This report reviews the literature on noise produced by wind turbines and its potential effects on human health. The reviewer concludes that published research and expert opinion by professional societies agree that there is no evidence that wind turbine sounds have any observable effects on human health.

Hubbard, Harvey H., and Kevin P. Sheppard. 1988. "Wind Turbine Acoustics Research: Bibliography with Selected Annotation." NASA Technical Memorandum 100528. Available online at http://docs.wind-watch.org/NASA-bibliography-annotation-Wind-turbine-acoustics-Hubbard-Shepherd-Jan-1988.pdf. Accessed on April 22, 2014.

This report is mentioned frequently by those who argue that wind turbines cause human health problems because of the noise they produce. The report lists 238 publications dealing with this topic, 18 of which are annotated. The report should be viewed with caution as turbine technology has changed significantly in the three decades since it was written.

Lantz, Eric, Ryan Wiser, and Maureen Hand. *IEA Wind Task 26: The Past and Future Cost of Wind Energy.* Golden, CO: National Renewable Energy Laboratory, May 2012. Available online at http://www.nrel.gov/docs/fy12osti/53510.pdf. Accessed on March 9, 2014.

As the title suggests, this report provides an excellent review of the role of wind energy in the nation's energy equation prior to 2012 with predictions and suggestions for its role into the future.

Lee, April, Owen Zinaman, and Jeffrey Logan. 2013. "Opportunities for Synergy between Natural Gas and Renewable Energy in the Electric Power and Transportation Sectors." National Renewable Energy Laboratory, December 2012. http://www.nrel.gov/docs/fy13osti/56324.pdf. Accessed on April 22, 2014.

Two projected "fuels of the future" are natural gas and wind power. This report explores a variety of ways in which the two fuels can be used in conjunction with each other and as complements to each other in transportation and electric power generation.

Merlin, Tracy, et al. 2013. "Systematic Review of the Human Health Effects of Wind Farms." National Health and Medical Research Council, Canberra. Available online at https://www.nhmrc.gov.au/_files_nhmrc/publications/attachments/eh54_systematic_review_of_the_human_health_effects_of_wind_farms_december_2013.pdf. Accessed on April 22, 2014.

This report was commissioned for the purpose of determining potential health effects as the result of noise, shadow flicker, and electromagnetic radiation produced by wind turbines. Reviewers attempted to explain the differences found between anecdotal reports of health effects and controlled scientific studies, which tended to show no such results. They concluded that these differences could often be explained by factors such as respondents' attitudes toward wind farms, visibility of turbines, financial considerations from the siting of turbines, community involvement in decision making on siting of turbines, age and design of turbines, and the nocebo effect.

Mulinazzi, Thomas, and Zhongquan Charlie Zheng. 2014. "Wind Farm Turbulence Impacts on General Aviation Airports in Kansas." Topeka, KS: Kansas Department of

Transportation. Available online at http://www.copanational.org/files/windfarms_kansas.pdf. Accessed on April 21, 2014.

This study was conducted to find out if wind farms have measurable effects on the flight of airplanes traveling in their vicinity. Researchers found that such effects can be identified, especially among smaller planes flying into and out of small airports close to the wind farms.

Myszewski, Margaret A., and Merryl Alber. 2013. "A Survey of State Regulation of Offshore Wind Facilities." Report prepared by the Georgia Coastal Research Council, University of Georgia, Athens, Georgia, for the Georgia Department of Natural Resources, Coastal Resources Division.

This report reviews the regulation of offshore wind power facilities under federal law and under the law and regulations of eleven states bordering the Atlantic Ocean.

National Research Council. 2013. *Worker Health and Safety on Offshore Wind Farms.* Washington, DC: The National Academies Press.

This volume reports on a study by the Marine Board of the National Research Council about possible health and safety effects on workers who are or might be employed on offshore wind production facilities.

National Research Council. Committee on Environmental Impacts of Wind-Energy Projects. 2007. *Environmental Impacts of Wind-Energy Projects.* Washington, DC: National Academies Press.

This report is divided into four major parts. The first part provides a context for the findings reported in the document, including the status of wind power projects in the United States and around the world, along with a discussion of the potential benefits of wind power. The second part focuses on the ecological effects of wind farms, especially on birds and bats. The third part deals with the whole range of potential effects on humans, ranging from physical effects to emotional effects. The fourth part of the report synthesizes the results from the first three parts to provide some suggestions and guidelines for the siting and operation of wind facilities.

"Renewable Energy: Wind Power's Contribution to Electric Power Generation and Impact on Farms and Rural Communities." 2004. Washington, DC: General Accountability Office.

This report describes the benefits to individual farmers and rural communities in the leasing of land for wind farm projects and for the construction of distributed wind systems for the production of electricity for farm uses.

Shepherd, Dennis G. 1990. "Historical Development of the Windmill." National Aeronautics and Space Administration. Available online at http://wind.nrel.gov/public/library/shepherd.pdf. Accessed on March 15, 2014.

This report, prepared for the U.S. National Aeronautics and Space Administration (NASA), is one of the most concise, readable, and complete histories of the development of windmills currently available.

Street, Thomas. 2008. "Climate Change, Offshore Wind Power, and the Coastal Zone Management Act." Available online at http://www.researchgate.net/publication/224556796_Climate_change_offshore_wind_power_and_the_coastal_zone_management_act. Accessed on March 6, 2014.

This report is based on the assumption that global climate change necessitates the search for alternative sources of energy that do not depend on the combustion of fossil fuels. The author reviews the potential of offshore wind projects for meeting this demand and the regulatory restrictions placed on the development of this technology by the Coastal Zone Management Act of 1972.

"20% Wind Energy by 2030: Increasing Wind Energy's Contribution to U.S. Electricity Supply." DOE/GO-102008-2567,

July 2008. Available online at http://www1.eere.energy.gov/wind/pdfs/41869.pdf. Accessed on April 18, 2014.

This report considers the factors involved in the nation's efforts to increase its wind power output to the point that it will be able to produce 20 percent of the nation's electricity by the year 2030. The report is a treasure trove of data, information, and historical information about the role of wind power in the U.S. energy equation.

Wind Energy: Additional Actions Could Help Ensure Effective Use of Federal Financial Support: Report to Congressional Requesters. Washington, DC: U.S. Government Accountability Office.

This report was prepared in 2013 by the U.S. Government Accountability Office in response to a request from three members of the House of Representatives to prepare a description of the then-current status of wind power in the United States. The agency's research found that efforts to develop wind power by the federal government were distributed among a wide variety of agencies with frequent overlap of objectives and programs. The GAO recommended changes that could be made to make research on wind power more efficient and effective.

Online Sources

"Advanced Wind Energy: Airborne Wind Turbines." Makani. http://www.google.com/makani/. Accessed on April 19, 2014.

This Web page is the Web site for a company that designs and makes high-altitude wind turbine systems, a breakthrough in wind technology that some people believe may be an important step forward in the development of wind power systems.

"Altaeros Energies Poised to Break World Record." Altaeros Energies. http://www.altaerosenergies.com/index.html. Accessed on April 18, 2014.

This Web site is the home page for a company that plans to build wind turbines that float at significant heights above sea level. This technology is intended to take advantage of the significantly higher wind speeds that are available at altitudes greater than those at which turbines are normally sited.

Barnard, Mike. "Airborne Wind Energy: It's All Platypuses instead of Cheetahs." Clean Technica. http://cleantechnica.com/2014/03/03/airborne-wind-energy-platypuses-instead-cheetahs/. Accessed on April 19, 2014.

This Web site provides what is probably the most complete and easy-to-understand introduction to high-altitude wind turbine technology currently available on the Internet. The article provides a history of the technology, current commercial efforts to bring it to reality in today's wind market, and problems entrepreneurs face in making this step happen.

Barnard, Mike. "Wind Energy Health Issues Fail the Test of Law Repeatedly." Barnard on Wind. http://barnardonwind.com/2014/02/23/wind-energy-health-concerns-fail-the-test-of-law-repeatedly/. Accessed on March 8, 2014.

Barnard reports on an extensive review of court cases from the United States, Canada, Australia, New Zealand, and other nations dealing with the purported effects of wind turbine operation on human health. He finds that in forty-six of the forty-seven cases reviewed, courts found no basis for believing that wind turbine operation has deleterious effects on human health.

"Bibliography for NEWEEP Webinar: Impact of Wind Power Projects on Residential Property Values." http://

www.windpoweringamerica.gov/newengland/pdfs/2010/webinar_neweep_property_values_bibliography.pdf. Accessed on April 22, 2014.

This bibliography was prepared for a Webinar on the effect of wind power facilities on property values. Although somewhat dated, the results provide a very useful overview of this issue.

"Bibliography for NEWEEP Webinar: Wind Turbines Noise and Health: Fact vs. Fiction Simulcast." http://www.windpoweringamerica.gov/newengland/pdfs/2010/webinar_neweep_wind_noise_health_bibliography.pdf. Accessed on April 22, 2014.

This bibliography contains an extensive listing of print and electronic references on the topic of the effects of wind turbines on human health.

"Birds and Bird Habitats: Guidelines for Wind Power Projects." Ontario Ministry of Natural Resources. December 2011. http://www.mnr.gov.on.ca/stdprodconsume/groups/lr/@mnr/@renewable/documents/document/stdprod_071273.pdf. Accessed on April 22, 2014.

This document provides a detailed and exhaustive review of the threat posed to birds by wind farms and outlines the steps that need to be taken by wind companies to reduce the risk posed by their machines to birds.

"Birds and Wind Development." American Bird Conservancy. http://www.abcbirds.org/abcprograms/policy/collisions/wind_developments.html. Accessed on April 22, 2014.

This Web page provides a detailed analysis of the threat posed by wind farms to birds with an interactive map that shows regions in the United States that pose greater or less threats to birds from wind facilities.

Curry, Andrew. "Hope for Stemming Wind Energy's Toll on Bats." National Geographic Daily News. http://

news.nationalgeographic.com/news/2010/09/100915-energy-wind-bats/. Accessed on April 21, 2014.

This online article describes a method for reducing the number of bats killed by wind turbines by reducing their operations during periods when the bats are most active.

Database of State Incentives for Renewables & Efficiency (DSIRE). http://www.dsireusa.org/. Accessed on March 22, 2014.

This Web site contains a complete summary of all state, federal, and local programs for the promotion of research and development on all forms of renewable energy. It is an invaluable resource on the range of programs that the federal government and states have developed to encourage wind energy, along with other forms of alternative energy.

"EERE's National Mission." http://www1.eere.energy.gov/office_eere/pdfs/electricity_stakeholder_pres_0513.pdf. Accessed on March 9, 2014.

This page discusses the role of alternative energy sources, including wind energy, in the FY 2014 federal budget proposed by President Barack Obama. It provides an excellent summary of the attention paid to wind energy in the past, along with a look at its potential in the near future of American energy policy.

"The Energy Bible." http://energybible.com/default.html. Accessed on March 22, 2014.

This Web site contains an almost limitless amount of information on every type and aspect of renewable energy, including solar, geothermal, wind, water, and bio energy. Information on wind energy is found at http://energybible.com/wind_energy/index.html. The information is not only general in nature but is also aimed at potential practical applications by individuals and communities.

Fagiano, L., and M. Milanese. "Airborne Wind Energy: An Overview." *Proceedings of the American Control Conference.* June 2012. 3132–3143. Available online at http://ieeexplore.ieee.org/xpl/articleDetails.jsp?tp=&arnumber=6314801&url=http%3A%2F%2Fieeexplore.ieee.org%2Fiel5%2F6297579%2F6314593%2F06314801.pdf%3Farnumber%3D6314801. Accessed on April 22, 2014.

The authors present a general introduction to the technology of airborne wind turbines.

Felker, Fort F. "Engineering Challenges of Airborne Wind Technology."
[Golden, Colo.]: National Renewable Energy Laboratory [2010]. http://www.nrel.gov/wind/pdfs/49409.pdf. Accessed on April 21, 2014.

This government publication presents a good general introduction to the topic of airborne wind systems with some of the problems they face in order to become a practical reality.

Flavin, Ron. "The US Department of Energy's 2014 Budget Request: Implications for Renewable Energy Funding." Renewable Energy.World.com. http://www.renewableenergyworld.com/rea/news/article/2014/03/the-us-department-of-energys-2014-budget-request-implications-for-renewable-energy-funding. Accessed on March 9, 2014.

This commentator provides an excellent review of the priorities for renewable energy, including wind energy, in President Barack Obama's proposed 2014 budget.

Hand, M. Maureen. 2013. *Tracking the Cost of Wind Energy.* Boulder, CO: North American Wind Energy Academy 2013 Symposium. http://www.nrel.gov/docs/fy14osti/60237.pdf. Accessed on March 20, 2014.

This excellent presentation reviews the elements involved in the pricing of wind energy, summarizes changes in the cost of wind power over the preceding half century, and outlines possible trends in the price of wind energy in the future.

Hoste, Graeme R. G., Michael J. Dvorak, and Mark Z. Jacobson. 2011. "Matching Hourly and Peak Demand by Combining Different Renewable Energy Sources: A Case Study for California in 2020." http://www.stanford.edu/group/efmh/jacobson/Articles/I/CombiningRenew/HosteFinalDraft. Accessed on April 22, 2014.

The authors suggest that electrical demands can be met solely by renewable sources provided that grid characteristics and other relevant factors can be modified so that energy from a variety of sources can be drawn upon in combination to meet changing demands in the city. They offer the current electrical system in California as providing an example as to how this objective can be achieved.

"How Windpower Works." How Stuff Works. http://science.howstuffworks.com/environmental/green-science/wind-power8.htm. Accessed on March 28, 2014.

This Web site provides some basic information on the role that tax credits and other financial incentives work to promote the development of wind power systems.

"Illinois Windmills." http://www.illinoiswindmills.org/index. Accessed on March 17, 2014.

In spite of its name, this Web site provides an excellent introduction to the subject of windmills in general and to the history of American windmills in particular.

"Illustrated History of Wind Power Development." http://telosnet.com/wind/early.html. Accessed on February 26, 2014.

This excellent Web site provides a detailed description of the development of wind power from its earliest days in Persia (about the sixth century CE) to the present day in two sections, early history through 1875 and 1875 to the twentieth century. It also includes information on a number of other wind-related topics, such as the arguments in favor of using wind power, government wind power programs, recent market developments, and the future of wind power.

Jannis, Heilmann. "The Technical and Economic Potential of Airborne Wind Energy." http://dspace.library.uu.nl/handle/1874/258716. Accessed on April 22, 2014.

This thesis explores methods for determining the potential of airborne wind systems as sources of electrical power in a variety of physical settings.

Johnson, Gregory D., Edward B. Arnett, and Cris D. Hein. 2014. "A Bibliography of Bat Fatality, Activity, and Interactions with Wind Turbines." http://www.batsandwind.org/pdf/Bibliography%20August%202013v2.pdf. Accessed on April 22, 2014.

This bibliography consists of the about 250 published and unpublished reports on the interactions of bats with wind turbines and other structures. The references are not annotated.

Langman, Jimmy. "Chilean Wind Farm Faces Turbulence over Whales." National Geographic Daily News. http://news.nationalgeographic.com/news/energy/2011/11/111129-chile-wind-farm-impact-on-the-blue-whale/. Accessed on April 21, 2014.

This story describes a developing conflict in Chile over plans to place a wind farm overshore near the Chilean coast, an area that includes a feeding ground for the endangered great blue whale.

Layton, Julia. "How Wind Power Works." How Stuff Works. http://science.howstuffworks.com/environmental/green-science/wind-power.htm. Accessed on March 28, 2014.

This article is especially useful because it covers a whole range of topics relating to wind power, including not only the technology and history of wind turbines but also the political and economic issues related to the topic.

Levitan, Dave. "High-Altitude Wind Energy: Huge Potential—and Hurdles." Environment 360. http://e360.yale.edu/feature/high_altitude_wind_energy_huge_potential_and_hurdles/2576/. Accessed on April 19, 2014.

High-altitude wind systems involve the lifting of wind turbines hundreds of meters above Earth's surface to take advantage of the stronger and more persistent winds found there. This article provides an excellent introduction to the technology involved in the methodology and discusses its probable chances of success as a commercial enterprise in the future.

Lynch, J. Peter. "Feed-in Tariffs: The Proven Road Not Taken . . . Why Not?" Principal Solar Institute. http://www.principalsolarinstitute.org/uploads/custom/3/_documents/Feed-In_Tariffs.pdf. Accessed on April 22, 2014.

The author argues that feed-in tariffs (FITs) are a strong motivator for the development of renewable sources of energy, such as wind power, but they have thus far not been extensively or wisely used for that purpose. He provides examples of the successful use of FITs in other countries and discusses why they have not yet been widely introduced into the United States.

Madsen, Kelly. "Adding Wind to the National Energy Equation." EPSCoR. https://iowaepscor-org.sws.iastate.edu/news/features/2013/green-grids. Accessed on April 21, 20914.

As the demand for and supply of energy from wind increases, questions are beginning to arise as to how this form of energy can be incorporated into the existing electrical grid and the overall energy system of the United States.

"Marshall County First to Ban Wind Farms." WNDU.com. http://www.wndu.com/home/headlines/Marshall-County-first-to-ban-wind-farms—208208491.html#.UZtUNsqv_B0. Accessed on April 21, 2014.

The Marshall County Board of Commissioners decided to ban the construction of wind farms in the county because the population density was just too great to accommodate the inconveniences and problems created by the presence of wind turbines. The decision was said to be the first such action by any county in Indiana and one of the very few such actions to be taken anywhere in the United States.

"The Next Generation of Airborne Wind Turbines." Energy Matters. http://www.energymatters.com.au/index.php?main_page=news_article&article_id=4242. Accessed on April 22, 2014.

This Web page provides an excellent introduction to the subject of airborne wind turbines and includes a video of the operation of one such machine.

Pearson, Sam. "Desert Storm: Battle Brews over Obama Renewable Energy Plan." National Geographic Daily News. http://news.nationalgeographic.com/news/energy/2013/07/130725-obama-renewable-energy-plan-public-lands/. Accessed on April 21, 2014.

Although the administration of President Barack Obama has continued to emphasize the importance of renewable energy, such as solar and wind, some of his historic political allies—environmentalists in particular—are expressing concern as to how these technologies are likely to impact endangered and threatened species and environments in the country.

Pierpont, Nina. Wind Turbine Syndrome. https://www.windturbinesyndrome.com/2014/is-big-wind-the-new-big-tobacco/. Accessed on March 8, 2014.

The author argues that the operation of industrial wind turbines is responsible for a host of human medical problems and that the industry, in attempting to withhold information about these problems, is mimicking the long-standing behavior of the tobacco industry in masking the health effects of their products.

"The *Real* Truth about Wind Energy." Sierra Club Canada. http://www.sierraclub.ca/sites/sierraclub.ca/files/wind_report_final_draft.pdf. Accessed on July 30, 2014.

This publication was written to counter claims by some individuals and organizations that wind turbines are responsible for a variety of human health problems. The report begin with the statement that "we are confident in saying there is no evidence of significant health effects that should prevent the further development and implementation of wind turbines, wind farms and wind energy." (emphasis in original)

"U.S. Fish and Wildlife Service Land-Based Wind Energy Guidelines." http://www.fws.gov/windenergy/docs/WEG_final.pdf. Accessed on April 22, 2014.

This publication provides guidelines for the construction of wind farms so as to reduce the threats they pose to the natural environment. The guidelines are divided into five tiers: preliminary site evaluation, site characterization, field studies of wildlife species present at the site, post-construction studies of potential impact on wildlife, and other post-construction studies.

"What Is Wind Energy?" Ammonit. http://www.ammonit.com/en/wind/wind-energy. Accessed on March 14, 2014.

This Web page is sponsored by the Ammonit company, which makes measurement instruments for wind and solar energy generation. It provides important basic information on a number of wind-related issues, such as a history of wind energy, pioneers in the development of wind instruments, and formulas for calculating the amount of wind energy that can be obtained from a wind device.

"Wind Power." Climate Warming Central. http://www.climatewarmingcentral.com/wind_page.html. Accessed on April 19, 2014.

This Web page is part of a Web site on global climate change. The page discussed the relevance of wind power to this issue. It reviews and evaluates some of the arguments against the use of wind power.

"Wind Power." SBC Energy Institute. http://www.sbc.slb.com/SBCInstitute/Publications/~/media/Files/SBC%20Energy%20Institute/SBC%20Energy%20Institute_Wind_Factbook_May%202013.ashx. Accessed on April 22, 2014.

This document is an excellent resource for most of the basic issues related to wind energy. It includes chapters on key concepts of wind power; status and future development; research, development, and demonstration; economics, financing, and key players; environment and social impacts; and grid integration.

"Wind Power Problems, Alleged Problems, and Objections." http://ramblingsdc.net/Australia/WindProblems.html#Power_surges. Accessed on April 19, 2014.

This Web site is an excellent source of the objections that have been raised with regard to the use of wind power and the evidence for and against each such claim. The presentation is well balanced and augmented with a number of useful references.

"Wind Power Today: Federal Wind Program Overview." U.S. Department of Energy. Energy Efficiency & Renewable Energy. http://www.nrel.gov/docs/fy06osti/39479.pdf. Accessed on April 18, 2014.

This publication provides an excellent overview of the U.S. program for the development of wind power, outlining a series of objectives for the role of wind power in meeting the nation's energy needs through 2016.

7 Chronology

Introduction

Humans have been using wind power for a variety of purposes for centuries. The chronology provided here lists some of the most important events in the history of wind energy.

ca. 5000 BCE The earliest record of wind power use by humans appears in the construction of sailing ships that are propelled by wind.

ca. 1700 BCE Archaeological records suggest that the Babylonian ruler Hammurabi develops plans for using wind power to operate a large-scale irrigation project.

ca. 200 BCE The first windmills designed for pumping water and grinding grain are established in China.

ca. 100 CE The Greek mathematician and natural philosopher Hero (also Heron) of Alexander invents the first known mechanical device to be driven by wind power.

ca. 400 CE Legend has it that Buddhists first begin using prayer wheels, small mechanical devices used during their regular devotions.

ca. 900 CE Piruz Nahavandi, a slave to the Persian caliph Umar (also Omar), is said to have designed the first vertical-axis windmill. The device becomes very popular, and by the

---

Nuclear power and wind energy are alternative forms of energy, both of which raise a number of difficult political, economic, social, environmental, and other issues. (Shutterstock)

tenth century, it is being widely used throughout the country, especially in the windy province of Seistan (also Sistan). Use of the windmill for lifting water and grinding grain soon spreads to the Middle East and Central Asia.

1180 First known windmill is constructed in Europe in the canon of Sauveur-le-Vicomte in France.

1218 The first documented evidence for the use of windmills in China occurs in the writings of statesman Yehlu Chhu-Tshai.

1745 English millwright Edmund Lee invents the fantail, a device for keeping a windmill pointed into the wind.

1833 German-born American utopian thinker John Adolphus Etzler proposes that devices be invented that can be lifted into the air, where they will be able to harvest the energy of the winds more efficiently than they do on the ground. His ideas are not realized in actual designs for more than 150 years.

1857 Connecticut mechanic Daniel Halladay invents the American windmill, a variation of earlier windmills in which the machine's sails or blades are mounted directly on a central pole.

1887 James Blyth, then professor at Anderson's College (Glasgow), now Strathcylde University, invents the first wind turbine for the production of electricity. One of his three designs is said to have provided his home with electricity for more than two decades.

1887 American inventor Charles F. Brush builds what he thinks is the first wind turbine for the production of electricity at his Ohio home. The device produces 12 kilowatts of power, which he uses to charge more than 400 batteries in the cellar of his home.

1888 American inventor LaVerne Noyes and engineer Thomas Perry invent the Aermotor windmill, a variation of earlier windmills that achieves great success because it makes use of the best available materials and engineering technology in its design.

1891 Danish inventor and educator Poul la Cour constructs an experimental wind turbine at Askov Folk High School to demonstrate the feasibility of using wind energy to provide electricity to rural areas of the country. The turbine runs until 1958.

1902 The largest pure sailing ship ever built, the *Thomas W. Lawson*, is launched from Quincy, Massachusetts.

1903 La Cour founds the Danish Wind Electricity Company for the purpose of providing information about the potential value of wind energy and for training workers in the field.

1919 German physicist Albert Betz derives a mathematical formula for determining the amount of kinetic energy that can be obtained from wind.

1919 Danish engineers Johannes Jensen and Poul Vinding design and patent a new type of windmill that they call the Agricco. Its blades operated like the wings of an airplane, capable of turning into different positions to adjust to the airflow. The Agricco also rotated to face into the wind automatically.

1920 Betz derives a formula that shows the maximum amount of kinetic energy that can be derived from a wind turbine, in contrast to the theoretical amount that can be obtained from such a source.

1920 German engineer and inventor Kurt Bilau uses his knowledge of aerodynamic principles to invent a new type of windmill capable of producing electricity from wind. He calls his invention the Ventimotor, a machine that looks very much like a traditional four-bladed horizontal windmill.

1922 Joe and Marcellus Jacobs build a wind turbine on their ranch near Vida, Montana, for the production of electricity for their property. The endeavor is so successful that neighbors ask the Jacobs brothers to construct similar structures for their own ranches, leading to the formation of the oldest wind power company in the United States, Jacobs Wind Electric Company, a corporation still in existence.

1922 Finnish engineer Sigurd Johannes Savonius invents a vertical-axis wind turbine for the generation of electricity, a device that remains in use today.

1927 Inventor brothers John and Gerhard Albers, of Cherokee, Iowa, construct a two-bladed wind turbine called the Wincharger for use in generating electricity for small facilities, such as a home or farm.

1931 French aeronautical engineer Georges Jean Marie Darrieus designs a vertical-axis wind turbine for the generation of electrical power that becomes widely popular and remains so today.

1931 German inventor Hermann Honnef designs a mammoth wind turbine for use in Berlin consisting of three (later increased to five) rotors with diameters of 160 meters, each mounted on a tower 500 meters high. Honnef believed that a machine of such height would be able to take advantage of greater wind speeds at higher altitudes above the ground. The turbine was actually built but never put into operation because of disruptions by World War II.

1935 The Rural Electrification Administration (REA) is created to bring electricity to rural areas of the United States, where it had previously not been generally available. The work of the REA severely depresses the demand for windmills in rural America.

1941 The first megawatt wind turbine, the Smith–Putnam wind turbine, is built on Grandpa's Knob, New Hampshire. The machine functions only briefly before breaking down.

1942 German engineer and inventor Franz Kleinhenz designs a wind turbine very similar to that first proposed by Honnef (see 1931) with a height of 250 meters, a rotor diameter of 130 meters, and a power output of 10 megawatts.

Mid-1940s The F.L.Smidth Company begins building wind turbines called Aeromotors for commercial use in Germany. The company had lost all of its cement-making business as a result of the war and turned to the construction of wind turbines as a substitute for its earlier business.

1949 The Federal Republic of Germany creates the Studiengesellschaft Windkraft e.V. (Society for the Study of Wind Power) to explore the possibility of building wind turbines capable of meeting a significant fraction of the nation's electrical needs. The society establishes a competition for the design of a wind turbine capable of producing 100 kilowatts of energy at a wind speed of 8 meters per second.

1950 Scottish engineer John Brown constructs and tests one of the first wind turbines in the United Kingdom on Orkney Island. The pilot project is not a success, probably because of the high winds at the turbine site. (A film of the project is available online at http://www.orkneywind.co.uk/costa.html.)

Early 1950s The German construction firm of Allgaier begins large-scale construction of wind turbines for sale worldwide. The primary purpose for which the wind machines were used was irrigation, but many were also sold for the production of electricity. The first turbines produced for this purpose became available in 1953.

1956 Johannes Juul, a student of Poul la Cour, designs and builds the first wind turbine to produce alternative current (AC) rather than direct current (DC). The machine is built at Gedser, Denmark, where it operates for ten years without requiring any maintenance.

1973 The Organization of Arab Petroleum Exporting Countries (OAPEC) declares an oil embargo against Canada, Japan, the Netherlands, the United Kingdom, and the United States in retaliation for their support of Israel in the Yom Kippur War of that year. The embargo inspires renewed interest in the development of renewable forms of energy, such as wind energy, to replace the world's dependence on fossil fuel reserves.

1973 President Richard M. Nixon announces his plans for Project Independence, a program designed to make the United States independent of foreign energy resources by 1980.

Development of alternative and renewable energy sources is an important component of that program.

1974 The U.S. Congress passes the Department of Energy Reorganization Act, creating the Department of Energy and defining new national policies for the development of alternative forms of energy.

1974 The U.S. Congress passes the Solar Research, Development, and Demonstration Act of 1974, which creates the Solar Energy Research Institute (SERI), later re-designated as the National Renewable Energy Laboratory by President George H. W. Bush in 1991.

1974 The U.S. Wind Energy Program is created at NASA's Lewis Research Center (now the Glenn Research Center) in Sandusky, Ohio. The program runs until 1992. During its existence, the program builds and tests thirteen experimental wind turbines. See also 1987.

1974 The American Wind Energy Association is established in Detroit, Michigan.

1975 The U.S. Energy Research and Development Administration (ERDA) creates a federal wind program to promote the development of wind energy facilities in the United States. The program is taken over two years later by the newly created U.S. Department of Energy. The program continues to exist today.

1976 The U.S. Department of Energy begins collecting data on wind speed in a number of locations to determine potential sites (so-called candidate sites) for the eventual location of wind farms. The project identifies seventeen such sites.

1977 President Jimmy Carter announces that national efforts to become energy-independent constitute "the moral equivalent of war." He includes the development of alternative energy sources as an important element in that effort.

1978 The U.S. Congress passes and President Carter signs the National Energy Act of 1978. Two sections of the act having particular relevance to wind energy are the Public Utility Regulatory Policies Act and the Energy Tax Act.

1980 The first commercial wind farm is established at Crotched Mountain, New Hampshire, an ill-fated project that survives little more than a year.

1980 The Department of Energy repeats its search for wind power candidate sites (see 1976) and identifies an additional twenty such sites for possible location of wind farms.

1981 Construction begins on a wind farm at the Altamont Pass in California. The wind farm is eventually to become the largest wind farm in the world, with more than 5,000 wind turbines now in operation.

1983 Iowa passes the first state renewable energy mandate, requiring investor-owned utilities to generate at least 100 megawatts of energy from renewable resources by some set date. The law is not implemented or enforced, however, until fourteen years later in 1997.

1985 Convinced that wind energy can make an essential contribution to meeting the electricity needs of citizens of the European Union, the European Commission creates a research group known as Wind Energie Größe Anlagen (WEGA; Wind Energy for Large Systems). The research project produces no practical devices and is later labeled as an "experimental" program. When a second phase of the project is initiated in 1991, the first phase becomes known as WEGA I. *See also* 1991.

1987 Researchers at the Lewis Large Wind Turbine Program build the largest wind turbine in the world at the time, the MOD-5B. The turbine as a rotor diameter of almost 100 meters (330 feet) and is capable of producing 3.2 megawatts of energy.

1987 A wind turbine called Éole begins operation at the Cap-Chat wind farm on Quebec's Gaspé Peninsula. The turbine is still the tallest vertical-axis wind turbine in the world, with a height of 110 meters and a nameplate capacity of 3.8 megawatts.

1981–1989 President Ronald Reagan reverses the vast majority of programs for the support and encouragement of research and development on renewable energy.

1991 President George H. W. Bush redefines and broadens the responsibilities of the Solar Energy Research Institute, renaming it the National Renewable Energy Laboratory (*see also* 1974).

1991 The world's first offshore wind farm is built near the village of Vindeby off the coast of Denmark's Lolland island.

1991 The first wind farm in the United Kingdom begins operation in Delabole, Cornwall. The facility consists of ten turbines that produce enough electrical energy to supply 2,700 homes.

1991 The European Commission launches WEGA II, the second phase of a program to develop wind turbines for large-scale applications in the European Union. The program lasts until 1997.

1992 The U.S. Congress passes, and President George H. W. Bush signs, the Energy Policy Act of 1992. The act provides for a 1.5 cent tax credit for the production of energy from wind and other renewable resources. The tax credit provision expires and then is renewed on a number of occasions in succeeding years.

1993 The first commercial wind farm in Canada is built at Cowley Ridge in southern Alberta.

1993 The Kenetech company develops one of the first commercially successful variable speed wind turbines. Prior to this time, the vast majority of wind turbines operated at the same speed, irrespective of the speed of the wind. Variable speed turbines are less expensive to build, easier to maintain, and less wasteful of energy.

1993 The National Wind Technology Center is established at the National Renewable Energy Laboratory in Boulder, Colorado.

1994 A number of groups interested in wind energy form the National Wind Coordinating Collaborative to focus on issues of shared interests, especially the effects of wind power generation on the natural environment.

1995 The Federal Energy Regulatory Commission rules that the state of California may not require electrical utilities to buy electricity from small wind turbine facilities at prices higher than those established by the so-called *avoided costs* rule.

2001 The Cape Wind company applies for permission to build a large wind farm in Nantucket Sound. It would be the first offshore wind facility in the United States.

2007 The U.S. Department of Interior establishes the Wind Turbine Guidelines Advisory Committee (WTGAC) to make recommendations on regulations for the siting of wind projects. The committee is to include representatives from industry, environmental nongovernmental organizations, state wildlife agencies, and tribes.

2008 The American Wind Wildlife Institute is formed to consider issues relating to the effect of wind turbines on wildlife. The organization consists of about twenty wildlife and wind energy groups, such as the Sierra Club, the Union of Concerned Scientists, BP Wind Energy, the Audubon Society, Duke Energy Renewables, and First Wind.

2009 The world's first commercial deep-sea floating wind turbine, called Hywind, is set into place in the Åmøy Fjord near Stavanger, Norway.

2012 The U.S. Fish and Wildlife Service publishes "Land-Based Wind Energy Guidelines," a collection of suggestions about the siting of wind facilities in such a way as to reduce potential damage to the natural environment.

2013 A group of about two dozen conservative political and lobbying groups petition the U.S. Congress to allow the tax credit for wind energy to expire.

2013 Construction begins on the first offshore wind farm in the United States by the Cape Wind company's wind farm in Nantucket Sound.

2013 The U.S. Congress ignores pressures to do otherwise and acts to renew the tax credit for the production of wind energy and energy produced by other renewable sources.

2013 A Vestas V164 wind turbine 220 meters tall is installed at Østerild, Denmark. It is the tallest wind turbine tower in the world at the time of installation. The turbine also has the largest nameplate capacity, 8.0 megawatts.

2013 A company called Altaeros Energies announces a technology by which wind turbines can be lifted into the atmosphere, where they are able to harvest wind energy more efficiently than they are capable of doing on the grounds. The technology is a realization of a proposal made nearly 200 years earlier by futurist John Etzler.

Glossary

Introduction

A number of specialized terms are used in discussing the technology, economics, politics, and other applications of wind power systems. This chapter lists a number of the most important of those terms with their most common definitions.

alternative energy Usually any form of energy other than coal, oil, and natural gas. Often used interchangeably with *renewable energy, q.v.*

ambient conditions Environmental conditions (e.g., temperature and pressure) at a particular location and a given time.

American windmill. *See* wind engine.

anemometer A device used to measure wind speed.

base load (also **baseload**) The minimum amount of power a generating plant can supply to consumers. The term was invented to describe conventional (fossil-fueled) power plants and has little or no precise meaning for wind power facilities.

Betz limit The maximum theoretical power that a wind turbine can capture from the wind. The Betz limit is equal to 59.3 percent of the wind energy.

candidate site A geographical location whose wind characteristics make it a likely candidate for the construction of a wind farm.

capacity factor A measure of the actual amount of electrical output from a wind turbine compared to its maximum theoretical output.

coefficient of performance The ratio of the power captured by the rotor of the wind turbine divided by the total power available in the wind just before it interacts with the rotor of the turbine.

cut-in speed The wind speed at which a turbine's blades begins to rotate and the turbine begins to produce electricity.

cut-out speed The wind speed at which a turbine's blades stop rotating and the turbine stops producing electricity.

decibel (dB) A unit of sound intensity.

dispatch ability The ability of an electrical generating system to respond to demand for current; the rate at which the facility can "power up" to begin full functioning.

distributed generation A form of energy generation by a number of small facilities located close to consumers who will use that energy. The term is used in contrast to utility generation by large companies that ship their products to consumers at greater distances from the production site. Distributed generation also refers primarily to low-power facilities, generally less than 60 kilovolts in capacity.

downwind turbine A turbine whose hub and blades are faced away from the wind direction.

energy equation A term that refers to that mix of energy resources, such as coal, oil, natural gas, nuclear power, and renewable energy resources, on which a nation, a region, or the entire world depends.

fantail A device attached to a windmill designed to keep the machine's sails pointed into the wind.

feed-in tariff A renewable energy policy in which producers of electricity from renewable sources are guaranteed payments for the electricity they produce and access to the grid, generally by way of relatively long-term (15–20 years) contracts.

furling Any device or procedure that turns a wind turbine rotor at an angle to the wind when wind velocities become dangerously high. Furling allows the turbine to continue operating without the risk of physical damage from the wind.

gigawatt One billion watts of power.

green pricing A practice followed by some utilities in which electricity produced from nonconventional sources (e.g., wind or solar power) is sold at a higher price than that charged for electricity from fossil fuel plants. The practice is viable because some consumers are willing to pay more for energy produced by more environmentally friendly methods.

grid The network of generators and transmission lines needed to produce electrical current and deliver that current to consumers.

horizontal-axis wind turbine A wind turbine in which the axis is horizontal to the ground, pointed into the wind.

hub The part of a wind turbine to which its blades are attached.

infrasound Acoustic oscillations whose frequency is less than that normally heard by humans, about 16 hertz.

installed capacity The maximum amount of energy that can be produced by a wind turbine when it is operating at full power.

kilowatt One thousand watts of power.

megawatt One million watts of power.

molinology The study of mechanical devices that are powered by water or wind energy.

nacelle The housing that holds the engine and associated mechanisms used in a variety of devices, including a wind turbine.

net metering A process by which the owner of a small wind system can sell back to an electric utility any excess electricity that it generates from its own system.

panemone windmill The name given to a windmill with a vertical axis and vertical sails attached to the axis by horizontal struts.

peak demand. *See* peak load.

peak load That period during which demand for electricity is at its greatest. Also known as *peak demand.*

petawatt One quadrillion watts of power.

post windmill A style of windmill in which the main elements of the mill are mounted onto a very heavy single pole that runs through the center of the machine.

power The flow of energy per unit of time.

power curve A graphical representation of the amount of power produced by a wind turbine of given design at various wind speeds.

power surge A sudden, unexpected flow of energy with the ability to damage devices through which an electric current is flowing.

production tax credit (PTC) A tax credit given to the producers of electricity based on the actual amount of energy produced in a year.

renewable energy Usually any form of energy obtained from some natural source, such as solar, wind, geothermal, and tidal energy. Often used interchangeably with the term *alternative energy, q.v.*

rotor The portion of a wind turbine that consists of its blades and the base (hub) to which they are attached.

rotor diameter The diameter of the circle swept out by the rotating blades of a wind turbine.

shut down wind speed The wind speed at which a turbine must be shut down because higher wind speeds will cause physical damage to the device.

spinning reserve A condition in which an energy-generating plant continues to operate but only at a low maintenance rate. Also known as *spinning standby.*

survival wind speed The highest wind speed that a wind turbine can withstand without experiencing mechanical damage to the machine.

terawatt One trillion watts of power.

tower windmill A type of windmill in which a cap rests on the top of a central pole, around which it rotates as the wind blows from different directions.

upwind turbine A wind turbine whose hub and blades are faced into the wind.

utility scale wind A large wind facility, usually with a capacity greater than 100 kilowatts, in which electricity that is produced at the facility is sold to some distant customers rather than used on or close to the generating site.

wind engine A uniquely American form of windmill in which the machine's sails or blades are mounted on a central pole.

wind monitoring system A collection of devices used to determine the characteristics of wind, including at least an anemometer (for determining wind speed) and a wind vane (for determining wind direction).

wind turbine A device for converting the mechanical energy of moving air directly into electrical energy.

Index

About the Author

David E. Newton holds an associate's degree in science from Grand Rapids (Michigan) Junior College, a BA in chemistry (with high distinction), an MA in education from the University of Michigan, and an EdD in science education from Harvard University. He is the author of more than 400 textbooks, encyclopedias, resource books, research manuals, laboratory manuals, trade books, and other educational materials. He taught mathematics, chemistry, and physical science in Grand Rapids, Michigan, for thirteen years; was professor of chemistry and physics at Salem State College in Massachusetts for fifteen years; and was adjunct professor in the College of Professional Studies at the University of San Francisco for ten years.

The author's previous books for ABC-CLIO include *Global Warming* (1993), *Gay and Lesbian Rights—A Resource Handbook* (1994, 2009), *The Ozone Dilemma* (1995), *Violence and the Mass Media* (1996), *Environmental Justice* (1996, 2009), *Encyclopedia of Cryptology* (1997), *Social Issues in Science and Technology: An Encyclopedia* (1999), *DNA Technology* (2009), and *Sexual Health* (2010). His other recent books include *Physics: Oryx Frontiers of Science Series* (2000), *Sick!* (4 volumes) (2000), *Science, Technology, and Society: The Impact of Science in the 19th Century* (2 volumes; 2001), *Encyclopedia of Fire* (2002), *Molecular Nanotechnology: Oryx Frontiers of Science Series* (2002), *Encyclopedia of Water* (2003), *Encyclopedia of Air* (2004), *The New Chemistry* (6 volumes; 2007), *Nuclear Power* (2005), *Stem Cell Research* (2006), *Latinos in the Sciences, Math, and Professions* (2007), and *DNA Evidence and Forensic Science* (2008). He has also been an updating and consulting editor on a number of books and reference works, including *Chemical Compounds* (2005), *Chemical Elements* (2006), *Encyclopedia of Endangered Species* (2006), *World of Mathematics* (2006), *World of Chemistry* (2006), *World of Health* (2006), *UXL Encyclopedia of Science* (2007), *Alternative Medicine* (2008), *Grzimek's Animal Life Encyclopedia* (2009), *Community Health* (2009), *Genetic Medicine* (2009), *The Gale Encyclopedia of Medicine* (2010–2011), *The Gale Encyclopedia of Alternative Medicine* (2013), *Discoveries in Modern Science: Exploration, Invention, and Technology* (2013–2014), and *Science in Context* (2013–2014).